Published by Thomas Telford Ltd, 40 Marsh Wall, London E14 9TP
www.thomastelford.com

Distributors for Thomas Telford books are
USA: ASCE Press, 1801 Alexander Bell Drive, Reston, VA 20191-4400, USA
Australia: DA Books and Journals, 648 Whitehorse Road, Mitcham 3132, Victoria

First published 2009

A catalogue record for this book is available from the British Library

ISBN: 978-0-277-3524-9

Typeset by Academic + Technical, Bristol
Printed and bound in Great Britain by MPG Books, Bodmin, Cornwall

Preface to the 6th edition

Civil engineering is 'the art of directing the great sources of power in Nature for the use and convenience of man' (Thomas Tredgold, 1788–1829). Its products are structures and systems for society, public services, commerce and industry. Civil engineering is thus central to the economy of the UK and other countries. It is also a major employer. The success of its projects are therefore vital to individuals, companies and governments.

The first edition of this guide was produced in 1963 when estimating books still gave items for 'haulage by horse and cart' and 'compaction by horse-drawn roller'. The following editions covered the substantial changes in the methods and equipment introduced into the field of civil engineering over a period of thirty or more years as the industry in the UK changed to meet the demands of promoters for better value for money and to comply with greater legal regulation of construction health, safety and effects on the environment. In the twelve years since the last edition, the construction industry has had to embrace the strictures of the Latham and Egan Reports[1] as well as taking on board Private Finance Initiatives and Public Private Partnerships.

Traditionally the promoter of a project employed a consulting engineer to design the project and then employed a separate

[1] Sir Michael Latham, *Constructing the team*, final report, HM Stationery Office, London, 1994
Sir John Egan, Rethinking Construction, HM Stationery Office, London, 1998
Sir John Egan, Accelerating Change, Construction Industry Council, 2002

contractor to construct it, supervised by the consulting engineer. Alternative arrangements are now increasingly used for the design and construction of projects large and small, and also for maintenance, demolition and improvement work.

This Sixth Edition of the guide remains an introduction to the traditional procedure in civil engineering while tackling the many alternatives that have developed over the last twelve years. The production of this edition has been entrusted to the Management Panel of the Institution of Civil Engineers, a body that is made up of members both of that Institution and those of the Institution of Civil Engineering Surveyors. As a result some of the best current practitioners in the civil engineering industry have contributed to respective chapters of this guide, bringing together a wealth of engineering and commercial expertise.

Defined words used in the text of this guide appear in *italic* letters.

Drafting Panel

Martin Kennard, Consultant
Derek Smith, Turner & Townsend Contract Services
James Piggott, IBM
Ian Newiss, Jacobs
Jerry Greenhalgh, Costain Limited
Jim McDermott, First Engineering
Elizabeth Bell, Carnell
Simon Grubb, Interserve Plc
Mike Battman, Gardiner & Theobald LLP
Martin Howe, Bevan Brittan LLP

Contents

1
Projects

Project cycle

Projects vary in scale and complexity, novelty urgency and duration, but typical of most of them is a cycle of work as indicated in Fig. 1. Each stage in the cycle is different in the nature, complexity and speed of activities and the type of resources employed.

The duration of the stages vary from project to project, sometimes with delays between one and the next. Figure 1 shows the common sequence of these stages. One activity does not have to be completed before the next is started. They can overlap, especially in the case of an urgent or fast-track project.

Initial proposal

The cycle starts with an initial proposal to meet a demand for the goods or services the project might produce. The proposal is often based upon a concept, engineering ideas, experience and records from previous projects, together with information from research indicating new possibilities. The relative importance of information from research, demand, experience and records depends upon the extent of novelty of the proposal and how far innovation will be required in its design, but all of these sources of information are always relevant to some extent.

Feasibility studies

If the first ideas indicate that the project may be economically attractive, the cycle proceeds to what are commonly known as *feasibility*

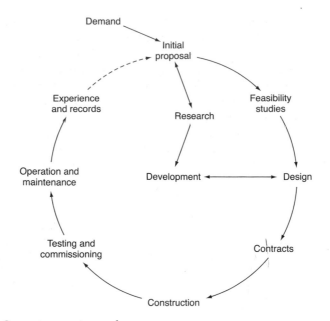

Fig. 1 *Stages in a project cycle*

studies to investigate a possible design and estimate its costs. Several alternative schemes likely to meet the expected demand are usually considered. In emergencies, this stage is omitted. If the project is urgent, little time is spent in trying to optimise the proposal. More commonly, alternatives have to be evaluated in order to decide whether to proceed and how best to do so in order to evaluate the risks and to achieve the *Promoter's** objectives within their budget.

The results of the first evaluation may be disappointing. If so, the proposal will be changed to try and meet the expected demand. The estimates of the demand may have to be reconsidered or made more precise and the first evaluation may also have to be reviewed because

---

* The 'Promoter' is used in the rest of this guide to indicate the particular promoter, *Employer* or *Client* for a project.

information used has changed during this process. This stage may therefore have to be repeated several times.

The investigation of schemes and their evaluation often proceed unsteadily because of the lack of information available upon which to make decisions. The results are bound to remain uncertain. Estimates of the possible costs of the project depend upon the risks of construction and the future prices of materials, *equipment* and labour resource. Estimates of the potential value of the project depend upon the probabilities of the result meeting the envisaged demand.

Estimates of a project's potential value are particularly uncertain when a capital project which will not earn money is being considered, for example, a project to improve safety at a road junction. Nevertheless a specification, budget and programme are normally decided, together with contingency, margins, programme and sources of funding. As the decisions made at this stage define the scope and standards of the project they form the basis of all that is to follow.

Project selection

The investigations and feasibility studies of a proposed project may take time. The conclusions have to be quite specific – selection or rejection of the proposed project. This determines the future of the project. Enough time and other resources should therefore be used to provide a valid basis for the decision. If the proposed project is rejected, it could be revived if new information is obtained on the demand, or a new design and other ideas that are more economic. Otherwise it is dead. The information used for the feasibility studies should therefore have been good enough for the best decision to be made. Similarly, if a proposal is selected, the information used should have been good enough to provide the start to a successful project.

If the project is selected and sanctioned the activities then move from assessing whether it should proceed to deciding how best it should be realised and to specifying what needs to be done. Figure 2 is a common graphical way of showing the sequence of work for a project. This is an example of a bar chart in which the lengths of the bars indicate the expected duration of each activity.

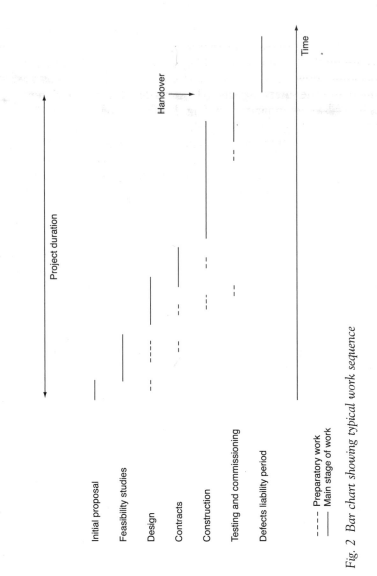

Fig. 2 Bar chart showing typical work sequence

Design

The decisions made early in the design almost entirely determine the quality, safety and cost, and therefore success of a project. Design ideas are often the start of possible projects. The main design stage of deciding how to use materials, the methods of construction required and the safety aspects to be considered to realise the project usually follows as indicated in Fig. 1. The products of design are drawings and a specification, although only a sketch may be needed for a very small project, repairs or maintenance work.

Intermediate stages of design may be needed to provide sufficient detail in order to check estimates of costs and provide a scheme for approval by the Promoter and *Statutory Authorities*. On a novel project, further research and development work may be needed to investigate new or risky problems before the project is continued.

On complex projects it is becoming increasingly more common to involve a *Contractor* in the design stage to better understand the economies that can be derived between the design and the construction processes.

Contracts

Figures 1 and 2 indicate that a *contract* for construction follows the completion of design. This is normal in the traditional procedure for civil engineering and building projects in the UK. Alternatively, only an outline design or performance requirement can be the basis of a contract and the chosen contractor then becomes responsible for design and construction. The various styles of contract governing relationships between the Promoter and the *Contractor* are discussed further in Chapter 6. *Consultants*, designers, *project managers*, suppliers and others are also usually employed under contracts,[*] some of them from the start of the investigation of a proposed project.

---

[*] Contracts to employ consultants are often called 'conditions of engagement' or 'service agreements'.

Construction

Construction usually requires larger numbers of people and a greater variety of activities than do the preceding stages. The cost per day rises sharply, as indicated in Fig. 3. So does the potential for waste and inefficiency. Construction therefore requires more detailed attention to its planning, organisation, health and safety, and costs. Demolition, substantial changes to existing structures and ground conditions require special care.

Most companies and public bodies who promote projects employ contractors from this stage forward to carry out the physical work on site. There is now a growing tendency by the Promoter, to seek *Early Contractor Involvement* (ECI) or to obtain construction expertise through the appointment of a *Construction Manager* through the design phase. Alternatively, contractors who take on the role of project promoters are normally fully responsible for design, for

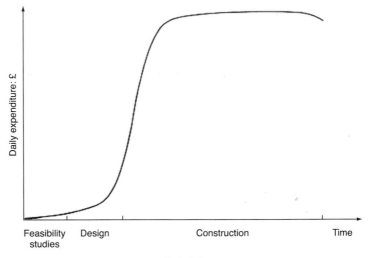

Fig. 3 Cost per day against project stage

example, when investing in the construction of a building for sale or financing an infrastructure project through a *Private Finance Initiative* (PFI). For virtually all construction the contractors in turn employ *specialist Contractors* and local *sub-contractors* to work on site and provide services, *plant*, materials and sub-systems.

Completion and handover

Different sections of a project can proceed at different speeds in both design and subsequent stages, but all must come together for commissioning and handing over of the completed facility for use. The investment should then be achieving all of the objectives of the project.

A subsequent proposal to alter the facility, replace its equipment and services, or decommission and demolish it is a separate project and should proceed through the appropriate sequence of investigations and decisions.

Objectives

Each stage in the cycle shown in Fig. 1 should be planned and managed in order to provide the best basis for all the following stages. The purpose of the whole sequence should be to produce a successful project. The following chapters of this guide describe how each stage of a project can be organised to achieve this result.

2

Promotion of a project

Project promoters

Legal status of promoters

The Promoters of civil engineering projects vary widely in their legal status. They include central, regional and local government and their agencies, limited liability companies, partnerships and business people who are sole traders. Other legally incorporated or unincorporated bodies may also be Promoters.

The advent of *Public Private Partnerships* (PPP) and Private Finance Initiative (PFI), has brought about Special Purpose Vehicles (SPV's) which are created from a mixture of Promoter types, including contractors.

Central government

Consultancy, construction and any other work done under contract with UK government departments and agencies is paid for out of funds provided by a vote of Parliament. Expenditure in excess of the voted amount, although it may be authorised by a department or agency, needs a supplementary vote. The government then authorises its departments and agencies to proceed with projects and studies for possible projects.

Contractors are not under any obligation to enquire whether or not the department or agency is contracting beyond the funds voted by Parliament. The government is bound by a contract made on its behalf by any of its departments or other agents which have the authority to enter into that contract.

It should be noted that in situations where projects are commissioned under either PPP or PFI arrangements, initial funding is made by the private sector. Government funding is made through payments against performance criteria during the operational phase of the contract, as opposed to the traditional payment for the construction as works progress.

Local government

The powers of local authorities in the UK are stated in their Charters and Acts of Parliament constituting them and under general or special Acts governing their procedures and applications of funds. Local authorities can enter into contracts and raise funds for payments due under these Acts and Charters. A contract validly entered into carries an implied undertaking that the authority possesses or will collect the requisite funds.

Local authorities are subject to the ordinary legal liabilities as to their powers to contract and their liability to be sued. Lack of funds is no defence to a legal action for payment. For example, a local authority's liability in a contract to pay for work will not be cancelled by the refusal of a government department to provide a grant or to authorise the expenditure.

Statutory boards and corporations

The powers of statutory boards and corporations are stated in the Acts of Parliament constituting them. The legal position of such bodies and of their officials is similar to that of local authorities. Former public utilities and nationalised industries which have been privatised are now incorporated companies.

Incorporated companies

Under English law an incorporated company can enter into contracts within the purposes of its memorandum of association or within the powers prescribed by any special Act of Parliament incorporating the company or any other Act granting it powers for a specific purpose.

Special Purpose Vehicles (SPV's)

Special Purpose Vehicle is a term used to describe an organisation formed to undertake a specific project (or series of projects) under PFI or PPP arrangements. It is usually formed of companies who have a common interest in the delivery of privately funded contracts and will typically include contractors and their strategic partners and stakeholders, for example key manufacturers, operators, designers and funders.

Typically, a new company will be created to act as the concession company for the project, with the stakeholders having agreed their share and contribution in the formation of the company. Although the shareholding capital in the SPV may be limited, the terms of an SPV concession will be such as to make the stakeholders jointly and severally liable to the Promoter for costs/losses in the event of default for whatever reason of one or more of the members of the SPV.

The contractual and legal arrangements are complex, particularly as the construction of the works is undertaken by various members of the SPV, but all have liabilities to it!

Other bodies

A person or persons who are not a trading corporation, for instance a partnership or a club committee, can be the Promoter of a project. A consultant or contractor invited to enter into a contract with such a Promoter should check the authority of one or more of the individuals involved to commit the rest in personal liability for the purpose of that contract and for making payments due under it. The safest course for other parties is to enter into a contract with a sufficient number of members of the body to ensure that their collective financial status is adequate.

Overseas

The legal status of companies and government bodies and the legal controls of construction vary from country to country, even within the Commonwealth or the European Union. Ascertaining the status and authority of Promoters and others overseas needs local expertise.[1] Experience and advice are also usually needed on the culture and customs of construction organisations and individuals.

The World Bank, United Nations Industrial Development Organisation, Asian Development Bank, European Investment Bank, European Bank for Reconstruction and Development and other funding agencies provide the finance for some civil engineering projects. Finance is also made available directly to poorer countries by more wealthy countries, usually through government departments, for example the Overseas Development Administration, London, and development authorities in the country where a project is to be built.

Study team

Many industrial companies and government bodies employ their own engineering and other professional staff on the first studies and evaluation of projects. They also employ consultants, as do most smaller or occasional Promoters. The Promoter usually enters into a contract (service agreement) for the consultant's services.

Many large traditional construction consulting engineering practices have become 'multi-disciplinary', being able to provide a range of services beyond their historical function in a more traditional role. Similarly, other *project management* consultancies have expanded the other way and are able to offer specialist engineering services as part of their portfolio. Such consultants are often badged as a 'one-stop shop', saving the Promoter from engaging a number of individual specialist consultancies.

Consulting engineer

In traditional civil engineering, in the UK, the Promoter employs a *consulting engineer* to investigate and report on a proposed project.

The consulting engineer may be a firm or an individual, depending on the size and risks of the project. The consulting engineer's role at this stage of a project is to provide engineering advice to enable the Promoter to assess its feasibility and the relative merits of various alternative schemes to meet his requirements. Other specialist consultants may be needed, in a team operating under a lead consultant.

Project sponsor

Many Promoters are organisations within which departments are employed on various functions such as the operation or maintenance of existing facilities, planning future needs, human resource management, finance, legal services and public relations. All these will have some interest in proposed projects, but they have different expertise and may have different objectives and priorities for each project.

A Promoter should therefore make one senior manager responsible for defining the objectives and priorities of a proposed project. It is the role of sponsor or 'project champion'. The role is not necessarily a separate one. It is logically part of the responsibilities of the manager who has the authority to ensure that sufficient resources are employed at this stage.

Promoter's (Project) Manager[*]

Projects are becoming more complex. There is a recognition that efficiencies can be made in the construction process through programme management, i.e. the undertaking of a series of individual projects under one umbrella and similarly, in the management of estate.

---

[*] The term Project Manager is used for many activities in today's construction industry. In this chapter the term Promoter's Manager will be used when referring to a direct employee of the Promoter or an external organisation employed to take on the role.

Political, commercial, legal and technical change is such that specialists are needed to advise on the best way to develop and promote particular types of project. In addition, the general public will expect to be consulted on major projects or projects which may affect the environment and quality of local life.

In these circumstances the Promoter's interests may be best served by the appointment of one internally employed person or an external organisation on a complex project, to plan and manage the project and co-ordinate relationships with other organisations. This role is particularly important for

- ensuring that the project objectives are drafted for agreement by the Promoter and relevant financial or statutory authorities
- advising on stakeholder relationships
- obtaining advice on the likely cost of the project, and possible sources of finance
- planning for site selection and acquisition
- planning for public consultation, a planning application and representation at public inquiries
- preparing and coordinating the project strategy or 'brief'
- developing the project procurement strategy
- planning for the appointment of the larger team and the systems, etc., needed for the life of the project.

Programme management encapsulates a wider role and introduces much greater emphasis on programming, logistics management and may by undertaken in the role of a delivery partner within an integrated project team. Such arrangements are now used for complex projects where the Promoter is not fully conversant in the procurement and delivery of construction works, for example an Airport Operator or Rail Franchisee.

In addition to the Promoter's Manager role, a programme manager may be responsible for

- overall control of the project delivery – milestone management
- the introduction and management of core processes
- budget control
- the balance of time, cost and quality considerations.

Estate management is often referred to as 'portfolio management' and may be undertaken by firms specialising in facilities management (FM) or by multi-disciplinary firms with FM capability. Promoters will generally set the requirements for repair and refurbishment programmes, undertaken as 'hard' FM, with maintenance of services, contracts for cleaning etc, known as 'soft' FM. Projects may be for one or the other or both.

Services provided by the Promoter's Manager or an outside organisation for portfolio management may include

- advising and preparing the project objectives schedules
- negotiations with stakeholders
- development of the planned maintenance regime
- arrangements for unplanned maintenance
- letting and management of various contracts
- project controls
- fiscal reporting.

Once a project has been sanctioned the Promoter's Manager, or the external organisation needs the authority of the Promoter to manage and control its design and construction.

The role of Promoter's Manager is also not necessarily a separate job from other work, depending upon the size, risks and importance of a project. Because of the time it demands it is usually separate from the role of the senior manager who is project sponsor. The Promoter's Manager may be the consulting engineer appointed to investigate and report on the proposed project. Alternatively the Promoter's Manager may be an employee of the Promoter, or a specialist company in project management.

Selection of the team

Selection of a Consulting Engineer

The selection of a Consulting Engineer should start with defining the expertise appropriate to the project. Corporate membership of the Institution of Civil Engineers (ICE) is recognised as the appropriate qualification for positions of responsibility in civil engineering. Corporate members of ICE comprise Members, who have appropriate education, training and experience, and Fellows, who are senior members who have held positions of major responsibility on important engineering work for some years. All are subject to the Institution's by-laws, regulations and rules of professional conduct. Consulting engineers traditionally practised in partnerships, but increasingly these operate as limited liability companies, which may be multi-disciplinary.

Advice for Promoters regarding suitable consulting engineers and their methods of engagement and working can be obtained from the Association of Consulting Engineers. A formal agreement (a contract) should be completed between Promoter and Consulting Engineer which sets out the duties and responsibilities of each party and the fees and expenses to be paid. The terms of employment of a consulting engineer need to be consistent with those of others employed on a project, for instance the Promoter's Manager, where the Promoter's Manager is a separate organisation from that of the Promoter and contractors. Sets of model terms are available for the employment of most professions.*

Overseas

Engineers working overseas must be prepared to adapt to different customs, contract arrangements, professional standards and terms of employment. In some overseas countries an engineer may practise only if registered for that purpose in accordance with the laws of the country. Some governments specify that any foreign consulting engineer must form an association or partnership with a local firm or agent before being permitted to practise in their country.

---

* Model terms of engagement of consulting engineers and also promoter's manager are listed in Appendix A.

Selection of a Promoter's Manager

The Promoter's Manager should be appropriately qualified and have adequate experience of the type of project and the duties and responsibilities that the Promoter intends to assign to him as the manager of the project. Membership of the Association of Project Managers (MAPM) is the recognised base *qualification* for project managers. A Promoter's Manager for a civil engineering project usually needs to have the expertise to advise on risk management, project planning, contracts and organising design and construction. The individual may also need the financial expertise to advise the Promoter on the expected whole-life cost of the project, cash flow and alternative sources of finance, or have access to this from others in the organisation. An understanding of other professions may therefore be important for the Promoter's Manager, particularly during the process of setting up a project team to manage a large and complex project.

The selection of the Promoter's Manager, be it an individual or a specialist organisation should follow closely the process outlined for the selection of a consulting engineer. The Promoter can obtain advice on this from the Association of Consulting Engineers and the Association of Project Managers. Increasingly the promoters of civil engineering and other projects are employing promoter's managers whose competence has been certified by the Association of Project Managers or the equivalent organisations in other countries.

Selection of project team

The Promoter's Manager of a large or complex project will require engineering and other assistance. It is normal practice for the Promoter's Manager to select and appoint the project team, but in some instances the Promoter may appoint the team on the advice of a specialist project management organisation.

The selection and appointment of the team should be based on the same criteria as those applied in the selection of the consulting engineer, care being taken to appoint a team that will have the expertise and the resources needed for the project.

Preparation of brief

Responsibilities

An important task for the Promoter's Manager is to ensure that the Promoter defines the objectives for the project and agrees a project brief to guide the next stage of work. The brief should state

- the Promoter's objectives and priorities
- how consultants and other resources are to be employed
- an outline programme and budget, setting dates and cost targets for the investigations and design studies needed for the feasibility study.

The brief should be designed to guide the investigation and evaluation of alternative engineering schemes which appear on initial consideration to meet the Promoter's needs. An example of this would be a brief to investigate the provision of a bypass to a congested town. There may be several acceptable alternative routes and often within these routes alternatives such as a tunnel or bridge crossing to a river. Cost-benefit studies, risk analyses and environmental impact assessments for each alternative will help to concentrate the choice down to two or three alternatives which meet the criteria set by the Promoter.

Some investigative work may be needed to reach a decision on which of the alternatives merit further evaluation. Often a short study backed up by site visits and the use of already existing geological, hydrological and other information will be sufficient, together with the consulting engineer's experience on similar projects in the past.

Outline programme and budget

If a date for completion of the project has been set by the Promoter, the outline programme should show a time limit for reporting back

17

and for consequent decisions by the Promoter, dates for starting design, placing contracts and starting construction. The programme should also show any dates critical for financial and statutory approvals and agreements by others.

The basis is now established for the next stage, the investigation of the proposed project and reporting the results and recommendations to the Promoter.

Reference

1. LORAINE R. K. *Construction management in developing countries*, Thomas Telford, London, 1991

3

Feasibility studies

Introduction

Feasibility studies are commissioned by the Promoter or another stakeholder with a vested interest in seeing the project go ahead. They are usually used to confirm the need for the project, establish the objectives, identify any constraints, investigate the relative benefit of different options and make a recommendation for a preferred option.

The scope and rigour of a feasibility study depends on the size, complexity and risk associated with the project, as well as the amount of work that has been done previously. Studies for major projects, such as an international rail route or new airport, may go through several phases and last years, if not decades.

The study is normally conducted by a consultant or an in-house team.

Confirming the need

The reason the project is needed may be clear, for example, additional accommodation for an expanding company. Often, though, part of the feasibility study is to confirm this need enabling clear objectives to be set against which alternatives can be evaluated.

One way to demonstrate the project's need is to consider the 'do nothing' option. Understanding the consequences of carrying on with the current situation can often highlight exactly why the project is required.

Table 1 Consequences of the 'do nothing' approach

Need	Example	Consequence of doing nothing
Statutory obligations	A water company upgrading a sewage treatment works to meet water quality standards	Fines for non-compliance, loss of reputation
Meeting new demand	An energy company constructing a new power station	Inability to meet demand leading to loss of market share
Achieving a return on the investment made	A property developer refurbishing an office block	Depreciation of the property and reduction in potential lease value

Categories of project needs, examples, and the consequences of doing nothing include those listed in Table 1.

Whilst the consequences may be self-evident, they should be assessed objectively and quantified where appropriate. This will involve collation of the relevant information, which may be a substantial piece of work in its own right. Predicting future demand might be part of the feasibility study, or may make use of existing forecasts. In the examples above, the water company would need to have a clear idea of the current water quality in order to specify the correct treatment process. This would be established through laboratory analysis of a series of water samples. The developer Promoter investing in his property may predict the future rental potential of that particular office block by referring to market trend analysis or making comparison with similar refurbished properties in the area.

Quantifying the need for the project helps to forecast the benefits it is expected to achieve and establish a business case for carrying out the work.

Establish objectives

Once the need for completing the project is clear, a set of objectives for the project can be established. This determines the criteria against which the success of the project can be measured. These include budgetary, programme, scope, quality and performance objectives. It is important that the team producing the feasibility study are clear about what is being considered and manage the Promoter's expectations about what the project is expected to achieve.

Budgetary

The budget available for the project will often be the determining factor when considering how to achieve the project's objectives. This can depend on the mechanisms available for funding the project and the level of risk the Promoter is willing to take on. It is also important to consider any cash flow objectives, such as maintaining a positive balance.

Programme

The time by which the project needs to be completed is usually, but not always, a key objective. For instance, constructing a new stadium for an Olympic games has a clear end date, but the deadline for constructing a new airport runway may be less tangible and will relate to transport capacity in other locations and forecasts of future demand.

Scope

It is important to agree the scope of the project in broad terms at an early stage. This helps to avoid ambiguity later on, when it is much more costly to change what has already been built. Figure 4 shows the opportunity to avoid future costs by establishing scope clearly early on.

21

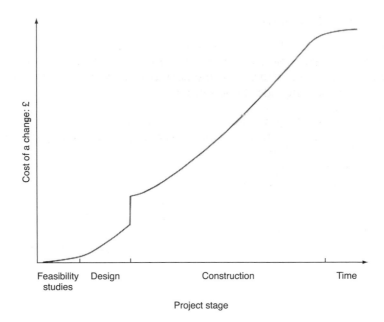

Fig. 4 Cost of change against project stage

Quality and performance

The project's service objectives can be considered in two, often inter-related, ways: quality and performance. Quality objectives define the standards expected and can often refer to industry standards in order to guarantee the adequacy of the completed work. An example would be the design of the structural elements of a bridge to recog-nised international standards. However, such standards do not always guarantee the project will meet the Promoter's needs. Perfor-mance objectives are used to define the parameters that are really important to the Promoter and are often related to capacity. For instance, performance objectives for the replacement of an ageing motorway bridge could include the ability to carry peak rush hour traffic for the next 50 years and have 30% less maintenance costs than the existing bridge.

Identify constraints

Part of the feasibility study is to define the range of factors that will affect the project. Establishing the constraints that exist allow uncertainty to be reduced and the project's risk to be managed. These factors typically include the following.

- Access to the site.
- Availability of utilities (water, sewerage, electricity, gas and telecommunications).
- Risk of flooding.
- Physical properties of the ground.
- Composition of waste expected from the site, e.g. from demolition and excavation, including any contaminated land.
- Suitability of existing infrastructure for reuse.
- Ecological/environmental constraints, including the presence of protected species.
- Whether the site is of archaeological interest.
- Rights of ownership to the land.
- Statutory approvals, e.g. planning permission.
- Unusual, complex or significant health and safety concerns.

These factors are established through a combination of desktop studies of existing records, site inspections, and specially commissioned investigations, including exploratory boreholes to establish soil conditions and environmental impact assessments.

As some of these items will have a long lead-time in relation to the time available for the feasibility study, starting to gather information is normally one of the first activities to be done. The scope of the investigations will depend on the time and budget available for the study, and the level of risk the Promoter is willing to take on before committing to further development of the project. Where a constraint has not been fully defined, the study should identify the next step that is required.

Stakeholders to the project should be identified during the feasibility study. A common way of establishing a communications strategy for the project is to map the level of parties' interest against the influence they possess. High interest, high influence parties constitute key stakeholders and early engagement with them may be appropriate during the feasibility study. This could include a public consultation.

Investigate and evaluate options

While the identification of constraints may limit the possible solutions, there may well be a number of ways in meeting the project's objectives, each with its own merits. During a feasibility study, each alternative under consideration should have a conceptual design produced to enable its impact on each of the constraints to be assessed and an appropriately robust cost to be estimated. The rigour and detail of the design and cost estimate will be proportional to the novelty of the project and the level of risk the Promoter is prepared to take on.

As well as its physical form, the method in which the project is delivered is an important consideration. The involvement of contractors that may carry out the work is often useful in determining the viability of different options, especially when considering methods of construction and buildability, which may have a significant bearing on the effectiveness of a solution. Involvement of contractors at this stage needs careful management of expectations and terms of engagement, in order that procurement of the main body of work is not prejudiced. There are likely to be a variety of methods available to procure the project, and their merits should also be considered during the feasibility study.

Early engagement of key stakeholders in identifying and evaluating options can provide them with ownership of their aspect of the project.

Evaluation of alternative options in a cost-benefit analysis should take into account the full range of project objectives in order to deliver best value. Where possible, measures should be given to intangible objectives, but it can be difficult to assign meaningful estimates and weighting to subjective or novel aspects. Relative comparison can be used in such cases to rank alternatives.

When comparing options, it is usual to consider whole life costs – the costs of maintaining and operating the project once complete, as well as the initial, capital, cost of building in the first place. Future costs (and returns) are usually discounted against current prices.

Make recommendation

Following evaluation of the alternative options, the study should be in a position to make a single recommendation. The recommended solution should be summarised in terms of the project's objectives in relation to its advantages over the other options. The proposals should be described in sufficient detail to allow its development by other parties. This would include an outline project strategy, including financing and procurement, programme, design and methods of construction. Outstanding areas of uncertainty should be identified, as well as the next steps required to mitigate risk. Although there may be the temptation to develop the solution in more detail at this stage, it is prudent to ensure the recommendation is agreed with before proceeding with further work in later stages of the project.

The results of the study will almost certainly be presented to a number of audiences. It can be useful to include a short summary for senior stakeholders who may not have the time to review the full detail. This should identify the need for the project, main objectives, principal constraints, options considered and why the recommended solution was selected.

4

Project strategy

Scope and purpose

A recommendation to proceed with a project should include proposals for the organisation and management of design, construction and other critical activities.

If the recommendation to proceed with the project is accepted, the Promoter's Manager should develop these proposals into a project strategy which states who is to do what and when. It should direct everyone to think about how the increasing scale of resources now needed should be organised, not just what has to be done.

Project management and control

Project Manager

From the start of design the Promoter's Manager or the externally appointed Project Manager[*] should be the prime channel of communication between the Promoter and all of the organisations or groups of people who will be employed by the Promoter to carry out the project. While others will input into the process and converse directly with the Promoter at various times, it is essential that the Project Manager has full knowledge of such actions and is in overall

---

[*] This guide assumes that the Promoter's Manager is an externally appointed Project Manager from this point forward.

control of the process. This is the lesson of many projects. The Project Manager should establish definitions of roles, authority, communication and reporting that bind everyone to the objectives.

Development of visual representations assists in clarifying the interaction between the Promoter and all the organisations that will deliver the project. Organisational structures (organograms), responsibility and accountability matrices (RACI diagrams) and communications trees are typical of the current requirements for major projects.

Project team

Once a project is to go ahead a project team should be formed with the expertise and resources to assist the Project Manager to plan and manage all of the remaining stages of the project through to hand-over.

If a design and construct, or Early Contractor Involvement (ECI) route has been selected, then the Contractor can be employed within the team from this stage.

Whether in the Promoter's or Contractor's organisation, the team should include staff that has recent experience of the construction, operation and maintenance of similar projects. Team knowledge of historical issues, for example, recurring problems with materials or processes, allows learning to be imported and generates a principle of continuous improvement, which in turn provides the opportunity for successive Projects to be more efficiently designed and managed.

The structure of the team should vary for each stage of the work, and will depend upon the size and complexity of the project and how much it can draw upon the resources of established departments within the Promoter's organisation, consultants and others.[1]

Flexibility is also required to accommodate sometimes significant and rapid change in the size and composition of the team. Work activities can be undertaken in sequence, where development of one aspect is dependent on the output from earlier work or in parallel where this is not the case. In general terms, parallel development will give rise to programme time savings, but imports some risk of integration of the various project elements. This risk is overcome by effective risk management as the project progresses.

Project controls

Project Control is a fundamental element of the Project Manager's toolkit.

In recent times, Project Control has become a sub-division of many Project Manager organisations, emphasising the importance of this subject in the successful delivery of Projects.

Most Project Control systems for major projects now utilise computer based standard and bespoke software programmes for various aspects of Project Control, particularly in relation to cost management, change management, risk management and documentation.

Except for minor or emergency projects the Project Manager should establish procedures for

- work breakdown, definition of authority, responsibilities and control of changes
- planning and progress monitoring of office and site manpower needs, project design and services, preparation and placing of contracts and other procurement, construction, critical sub-contracts, testing, commissioning and handover
- health and safety plans, statutory approvals, audits and reporting
- change management and the introduction of *variations*
- risk management
- cost estimating, cost management and trend analysis
- equipment and material ordering, inspection and delivery
- quality management plans
- audit traceability
- mapping and benchmarking of performance
- best practice and knowledge transfer capture and dissemination
- earned value reconciliation
- project management information
- reporting to the Project Manager

- reporting to the Promoter
- document records and control.

Planning

The purpose of planning the work for a project is to think ahead about what is needed to achieve the objectives of the project. Programmes and reports should include the amount of detail that will be needed by their intended users. Too little information can leave the team uncertain about what is wanted and what is happening. Too much detail can be counter-productive, as people will ignore it or at best only look for what they assume matters to them.

The planning of a project must allow time for the legal requirements of obtaining approval of design and construction by statutory authorities and for tendering procedures governed by EU Procurement Rules. Planning techniques should be used which are appropriate to the scale, urgency and risks of the project.[*] As a minimum, it is recommended that key milestones, trigger points and constraints are clearly identified and monitored.

With the advent of sophisticated programming and planning software, many complex projects now include specific stipulations for planning and programming software. Power project, MS Project and Primavera are currently the most readily recognised names.

Monitoring and reporting

All but minor routine projects need a system for monitoring progress and costs from the start of design to provide a basis for regular reviews of achievement and trends compared to programme and budget. The scope of the data that are to be presented in a report should be agreed with its users.

Attention to cost trends and probable outcomes is important on most projects, the exceptions being urgent work. For all projects speed in reporting costs is usually more valuable than accuracy, to

---

[*] See the publications on project planning and control listed in Appendix B.

enable action to be taken on trends. Quick data should be followed by accurate data and analysis of the causes of savings and extra costs in order to correct first impressions and improve the information available for estimating the costs of future projects.

Most monitoring now includes use of schedule performance indicators (SPIs) and cost performance indicators (CPIs) to allow data to be readily assimilated and trends identified.

Risk analysis and management

Project risks

The information used to decide whether to proceed with a project is inevitably based upon predictions and assumptions about the future conditions and costs which may affect its design, construction and completion. Political events, weather, the quality of design, unknown ground conditions, bankruptcies, plant failures, industrial relations, accidents, mistakes and criminal actions are all risks which may affect the progress, cost or economic value of proceeding. The identification and assessment of possible risks is therefore valuable at every stage through a project.

Not every party to a project will have the same view of a risk and how it should be managed. The Project Manager should therefore establish the Promoter's policy on risks and inform all the project team.

To assist this process, a comprehensive Risk Register should be developed, managed and maintained throughout the life of the project. Various software packages are available which provide tools for impacting the probability of events occurring and likely cost range into a consolidated risk profile. In common use are 'At Risk' and ARM, both use statistical sampling to predict the most likely outcome.

In modern parlance, risk should be placed with the party which is best equipped to manage it although there are still Promoters who prefer and pursue a policy of perceived total risk transfer. However, analysis of such procurement routes by the National Audit Office and others shows that this latter attitude is unlikely to provide the most cost effective outcome for the Promoter.

Risk management

This is a general title for a systematic procedure for identifying, analysing and deciding responses to risks.[2] It consists of three elements.

Risk identification

Experience and checklists are used to identify the sources of possible risks to the project, including physical, environmental, commercial, political, legal, financial, operational, technical, resourcing and logistical risks. Workshops are a useful forum for developing the initial risk register.

Risk analysis

The probability and potential effects of each risk are assessed, using systematic procedures such as the 'hurdle' method, sensitivity analysis, decision trees, probabilistic analysis and simulation. For risk analysis to be effective, regular risk reviews are an essential ingredient of Project Management. Shifting trends in probabilities or costs are good barometers of the success or otherwise of the project risks.

Risk response

For each risk, decisions are made as to whether to

- ignore it, if it is too unlikely or its potential effects trivial
- eliminate it, by changing the project
- transfer it, usually to an insurer or a construction contractor
- bear it, allow for its possible cost and other effects and manage it.

Health and safety management

Attention to health and safety is a professional and legal duty of every person involved with the promotion, design, construction and operation of a civil engineering project.[3–6] The construction

31

industry is one of the most dangerous. In addition to the human tragedies and waste, poor health and safety procedures at work and accidents affect morale and productivity and contractors and promoters lose money. All of this could be prevented by planning, training, incentives and good supervision.

The most important pieces of UK legislation are the Factories Acts and the Health and Safety at Work, etc Act, 1974 (Elizabeth II 1974 Chapter 37) – Reprinted 1991. Detailed requirements on sites are specified in the Construction Regulations. Requirements for attention to health and safety in design and construction are specified in the Construction (Design and Management) Regulations, 2007, (CDM). The legislation imposes criminal liabilities on everyone – promoters, consulting engineers, contractors, subcontractors and every individual as well as organisations – to anticipate and control risks to health, safety and welfare as far as is reasonably practicable.

The Construction (Design and Management) (CDM) Regulations

The CDM Regulations apply to construction, demolition and dismantling work. Under the Regulations, designers and contractors have to be selected on the basis that they are aware of their health and safety duties and will allocate adequate resources to them.

The Regulations impose on the Promoter (called the *Client* in these Regulations) the duty to select and appoint a *CDM Co-ordinator* and a *Principal Contractor* for a project. The relationships between the CDM Co-ordinator, Principal Contractor and Client are shown in Fig. 5.

The duties of the CDM Co-ordinator include but are not limited to

- ensuring that construction health and safety requirements are met in design
- ensuring cooperation between the designer and principal contractor

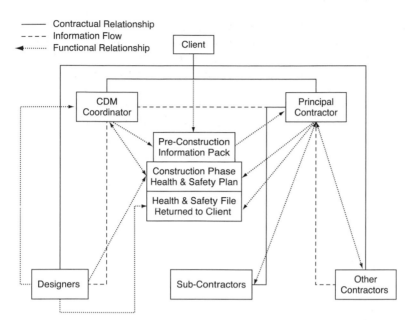

Fig. 5 Relationships of Planning Supervisor and Principal Contractor

- ensuring that a health and safety file is prepared for the project, is updated and is finally handed over to the Promoter.

The duties of the Principal Contractor include but are not limited to

- maintaining the health and safety file
- ensuring cooperation between all contractors on health and safety
- ensuring that everybody on-site complies with rules in the file.

Selection of the CDM Co-ordinator

The person appointed must be competent and have adequate resources. The Promoter may appoint the Project Manager, a consultant or any other person as the CDM Co-ordinator. A member of the Promoter's staff may be appointed.

Selection of the Principal Contractor

The Contractor appointed must be competent and have adequate resources. It can be the *Main Contractor* for constructing the project, a *management contractor* or a specialist in construction health and safety.

Quality management

Quality management is the process of ensuring that the project meets the Promoter's requirements for economic and safe quality of design, materials and construction.[7] Many Promoters only employ consultants and contractors who meet national and international standards for quality assurance (QA). Training, QA systems and detailed supervision on-site help greatly in achieving increased productivity and high standards of quality.

QA procedures are a systematic way of specifying actions which will give confidence that quality requirements will be achieved. These procedures should

- state how the quality standards required for a project are to be decided and communicated
- establish how and how far proposed methods of work are to be assessed and improved to achieve adequate standards of work
- specify how the records of standards achieved are to be compiled, coded and recorded
- include auditing of the effectiveness of the procedures.

QA procedures should be clear and brief, their purpose being to anticipate problems. QA should be required for a project only if the cost of establishing and applying it is expected to be less than the total cost risks if there were no QA system. Quality management policies and QA systems should therefore be reviewed regularly to see if they are cost-effective.

Promoter's procurement strategy

A project procurement strategy is the output of the process of deciding how to best deliver the project. It will marry the business case, project objectives and required deliverables (including considerations of time, cost and quality) and reflect the Promoter's approach to risk. It will reference the balance of the project team responsible, for the design, construction and other services needed for a project and may specify who should undertake all stages of work.

The procurement strategy adopted is of fundamental importance in governing the way in which the project will be undertaken, purchased, administered and delivered. It also exerts considerable influence over the project team's ability to achieve a successful balance between the costs, time and quality parameters.

Procurement strategies reflect fundamental differences in the allocation of risk and responsibilities between the parties; additionally the suitability of the different approaches have to be considered in relation to the specific nature of the individual project, as there is no one procurement route suitable for all.

In developing the optimum procurement route, the core business, strategic and project objectives and constraints must be considered, evaluated and prioritised.

Strategy choices

Business objectives and corporate policy drivers need to be considered in addition to the normal project drivers.

An optimum procurement strategy will evolve as a consequence of working through a staged evaluation process, which may include consideration of some or all of the following.

1. Should the Promoter employ consultants and contractors, or carry out all or some of the work with its own employees?
2. To what extent do the business drivers and project objectives influence the structure of the delivery team?

3. What is the envisaged shape of the project delivery team? If more than one organisation, are they to be employed sequentially or together?
4. Are enabling works desirable or necessary?
5. What are the legal considerations – Planning, EU Compliance?
6. Who is to be responsible for what? Who is to be responsible for defining objectives and priorities, design, quality, operating and maintenance decisions, health and safety studies, approvals, scheduling, procurement, construction, equipment installation, inspection, testing, commissioning and for managing each of these?
7. Who is to bear the risks of defining the project, investing in it, obtaining the necessary approvals, specifying performance, design risks, ground conditions, selecting sub-contractors, site productivity, mistakes and accidents?
8. What terms of payment will motivate all parties to achieve the Promoter's objectives – including payment for design, equipment, construction and services?
9. Is a form of incentive mechanism appropriate? If so, should it be time, cost or quality orientated – or a combination of each?

Some of the answers may be dictated by law, government policy or by financing bodies.

Contract strategy

The major contracting strategies are described below, and includes a commentary of the core functions and responsibilities undertaken by the Project Manager, common terminology and descriptions of the various contracting routes and contractual relationships and interaction.

The Project Manager

In Chapter 2, the term Promoter's Manager has been adopted to convey the role of the Promoters lead adviser. In the various contracting routes that are available under a procurement strategy, the role will require varying skill sets and the functions of the Project Manager will vary accordingly.

Traditionally in civil engineering in the UK the consulting engineer was the adviser to the Promoter from the inception of the project, as described in Chapter 2 and the route chosen was almost exclusively traditional contracting.

Under this route, when the project was sanctioned, a consulting engineer was appointed to be responsible for design, preparation of the construction contract and tender assessment. This role was then generally extended to the construction phase, where the consulting engineer became *the Engineer* named in the construction contract with powers and duties to supervise *the Works* and make decisions on design, construction and payment.

More recently, as it has become clear that 'one size fits all' is not a maxim that is appropriate to civil engineering projects, alternative forms of contracting have been used, particularly as a consequence of the Latham Principles, colloquially known as 'Constructing the Team' and Egan's 'Re Thinking Construction', in the mid-1990s.

In some contracting routes and forms of contract the title 'the Project Manager' is now used to mean the role of supervising the Works. Other titles including 'the Architect', 'Supervising Officer' and 'Contractor Administrator' are also used for similar roles in large building and some civil engineering contracts. The person's powers and duties vary from contract to contract, but in principle operate as indicated in Fig. 6.

However, the role of the Engineer with powers and duties to supervise the Works is still prevalent in international civil engineering contracts.

Contracts for construction work

It is usual to invite contractors to compete for a contract for construction work, in the expectation that they will plan to do the work efficiently and therefore at minimum cost. Unless the project is procured under an existing *framework* arrangement, EU tendering governance will apply for public works projects. The EU Procurement

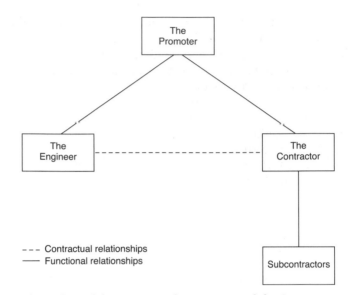

Fig. 6 Relationships of the Promoter, the Engineer and the Contractor

Regulations provide details of the required process and guidance in the application. It is usual to go through an *Expression of Interest* procedure and to then refine the tender invitation list by means of a transparent *Pre-Qualification* Process.

It should be noted that EU rules do not apply to private sector initiatives, although it will often be the case that frameworks exist within specific promoter bodies, or that a process akin to the public sector is desirable.

Selected contractors are invited to tender on an equal basis, competing on technical proposals, price and evidence of past performance. Competing on price and speed of starting work is common practice for sub-contracts. It is usual for the contractor who has submitted the most economically advantageous tender against the award criteria to be appointed.

Most contracts are for the construction of a defined project, although exceptions are *term* and *maintenance* contracts and those that are under a Private Finance Initiative (PFI) or Public/Private Partnership (PPP), all of which have the ability to include operation and maintenance as well as design and construction.

The commentary below covers the following contractual options and appropriate applications for

- traditional
- design and build
- develop and construct
- *Construction Management/Management Contracting*
- Term and Maintenance Contracts
- PPP/PFI.

Traditional Lump Sum Contracting

Traditional *lump sum* arrangements can be through a single or two stage tender process. Single-stage tendering will incorporate a full design prior to tendering for the construction works for a lump sum price. Early involvement of the Contractor is not obtained and the project team has less of an influence over the choice of sub-contractor, unless specifically named in the documentation. With two-stage tendering, the Contractor is brought on board early, tendering against an outline design and programme. The design is then completed with the Contractor's specialist input and second stage tendering is carried out on the basis of fully designed work packages.

Two-stage tendering is sometimes selected over single-stage when the works are complex, design team involvement is required during contractor tendering, and/or Early Contractor involvement is required.

Cost implication

Lump sum contracting provides a high degree of cost certainty providing that full design is achieved prior to tendering. Without this the Promoter is exposed to potential claims. Two-stage tendering can provide a more robust lump sum price on the basis that the Main Contractor is involved earlier in the process and interfaces with the design team prior to finalising their price.

Quality implication

Because design is retained by the Promoter's appointed designers, the desired level of quality can be consistently maintained throughout. There is however a limitation for designers to communicate directly with specialist suppliers and to effectively involve them at an early point in the project design process, which can be mitigated to some degree if the two-stage process is used.

Programme implication

In order to obtain full design prior to tendering, lump sum contracting requires a significant lead-in as usually no overlap occurs between design and construction. Early Contractor involvement can be achieved through a two-stage process.

Flexibility implication

While change can be incorporated under this route the tendency is for contractors to attempt to maximise rather than mitigate its effect. The Contractor's ability to do this is heightened by the fact that the Promoter and their advisors have no direct access to the Contractor's sub-contractors during the contract works.

Design and build

Under a design and build route, the Contractor is brought on board early and will initially tender against outline scheme requirements, for example a performance specification and drawings and draft contract; often these documents are colloquially referred to as 'Employers (Promoters) requirements'. These contracts are known by various descriptions, including Design-and-Construct, Package-Deal, All-in and *Turnkey*.

Once initially tendered, the design is worked up to a benchmark level (often against pre-determined criteria) where the Contractor and Promoter feel they are able to enter into contract. The Contractor may engage his own design team in this process or provision can be made for the Promoter's team involved in the initial, pre-tender phase, to be novated.

Cost implication

As with traditional lump sum contracting, design and build provides single point responsibility and a high degree of cost certainty, providing the Promoters Requirements are fully defined. Undefined Promoters Requirements will open the risk of subsequent claim. Risk transfer therefore needs to be fully considered and appropriately transferred.

Quality implication

Because the design responsibility is transferred to the Contractor's team, the Promoter loses direct control and hence quality can be compromised. While effective documentation mitigates this, the Contractor's focus may be on commercial issues rather than design. Options exist to either novate the design team to the Contractor or to largely complete the design prior to handing over to the Contractor to mitigate these considerations. If the latter option is chosen, it has to be recognised that some benefit of the Contractor's design input and commercial advantage is lost.

Programme implication

Design and build can achieve a faster start on site due to the overlapping of design and construction and through Early Contractor involvement. However, the programme must reflect sufficient time for defining the Promoter's requirements.

Procurement can be on a single- or two-stage basis, with the latter providing overall programme savings.

Flexibility implication

Design and build does not readily accommodate change in the Promoter's requirements. The Contractor, through his control of both design and sub-contractors, has a strong negotiating position and may look to maximise this effect for their financial gain.

Develop and construct

Develop and construct may be viewed as a derivative (or 'son of') design and build, but it recognises that most of the design work has already been completed. It therefore obviates some of the drawbacks of design and build, but reduces the Contractor's ability to bring their skills and knowledge to the design phase. As with design and build, design responsibility will be passed to the Contractor as well as the construction risk.

Management/package route – Construction Management/ Management Contracting

The Management/package route is usually considered under two main strategies, Construction Management (CM) and Management Contracting (MC). Both effectively use a series of specialist trade contractors to carry out the construction works, the difference arises in the contractual relationships.

Under CM, a Construction Manager is appointed to be the Contractor, albeit that his role is limited to the provision of site facilities and management and coordination of the construction works packages. The Construction Manager and the trade contractors are all appointed by the Promoter. In MC, it is the Management Contractor who engages the trade contractors and control of the supply chain is therefore passed to the Management Contractor.

The Promoter effectively employs the managing organisation as an extension to his own. Their duties vary from project to project, for instance in responsibilities for coordinating design and construction. The duties and powers of the managing organisation need to be stated consistently in its contract with the Promoter and in the contracts of the trade contractors.

In either case, reimbursement to the Construction Manager or Management Contractor is by way of a fee, plus cost or lump sum for the provision of site facilities.

The Promoter should possess a very good understanding of construction practices before embarking on this route.

Cost implication

A CM/MC procurement route overlaps design and construction, with the work being packaged and tenders achieved progressively as design packages are completed. As such, the Promoter does not have the benefit of a lump sum price prior to commitment (unless the route is adapted to accommodate this) and is thus exposed to financial risk. Mitigation is available through knowledgeable cost management, pre-market testing and early tendering of significant work packages. Because with CM, the Promoter is in direct contract with the supply chain, he is better able to control cost risk than with MC.

Quality implication

Quality is maintained as the Promoter retains control of the design team. Flexibility is also inherent because various work packages can be let as 'design and construct'. In addition, with CM the Promoter has the added advantage of a professional Construction Manager who is independent by virtue of having no financial interest beyond his management role.

Programme implication

The CM/MC routes provide major advantages in terms of programme as design and construction are overlapped. CM/MC also has the ability to re-plan and to incentivise Trade Contractors, particularly in order to mitigate delay or accommodate change, but the Promoter will carry the associated time and cost risks of disruption beyond the initial delay cause, whether of their making or not. The CM or MC is appointed prior to completion of full design to assist in programming, cost planning, buildability issues and to procure the works.

Flexibility implication

CM by its nature provides inherent flexibility in terms of managing and incorporating change. Because of the independence of the Construction Manager, the true consequences of change can be ascertained and re-planning of the works can be totally transparent. Direct access to trade contractors facilitates this process. MC is less flexible, although change can be incorporated more readily later in the process than with the traditional route on the basis that design and construction overlap.

Term and Maintenance Contracts

These types of contract are used when a Promoter has a need for an ongoing sequence of works, but where the extent is unclear. They may therefore be looked on as a contract for a series of Contracts (Term) or for dealing with planned and unplanned repairs and renewals (Maintenance). They may equally be seen as framework arrangements, where mini competitive tenders can be sought for specific works where a series of contractors already has an umbrella contract for Term and Maintenance works.

They are usually let for fixed periods of time, often 3–5 years, to provide opportunity for contractors to build efficient working practices, while giving the Promoter the opportunity to ensure pricing remains competitive at reasonable periods of time.

Contracts may be tendered against schedules of *rates* for various works or lump sum pricing/fee percentages.

Private Public Partnership (PPP)/Private Finance Initiative (PFI)

Comprehensive Contracts have become more popular with Promoters in the UK, and it is now common for both UK and overseas contractors to have to obtain the finance for a project.

Under PFI, the normal arrangement is for the finance of the Design and Construction phase to be provided by the private sector and to recover its cost plus profit over an operational period of typically 20–30 years, termed a concession. During this period the concessionaire receives the income from users, for instance from drivers using a toll road. At the end of the period, ownership of the project usually transfers to the government (or other initiating promoting organisation).

A variety of concession arrangements have been developed, with various names, but because the Contractors and their partners obtain the finance to construct the project, they are therefore the project Promoters.

The most commonly seen names for PFI arrangements are as follows.

- Build-Own-Operate-Transfer (BOOT) or can be described as Design, Build, Operate, Transfer (DBOT).
- Design-Build-Finance-Operate (DBFO).
- Design, Build, Operate, Maintain (DBFM).

The contractual arrangements are complex, requiring a head, concession, agreement, and contracts between the Special Purpose Vehicle (SPV) set up as the concessionaire and its key supply chain, most of whom are likely to be stakeholders in the SPV.

PFIs will be tendered against the ultimate (initiating) Promoter's Requirements, using performance specifications and requirements for the legal, contractual, financial and commercial model. Once a preferred bidder is selected, negotiations take place to seal the contract, termed commercial close. Because of the complexities involved in the process, this can often take a period of several months or even years.

The alternative to PFI is PPP, where under Public Private Partnership, there is a sharing of responsibilities for delivery of certain parts of a project. PPP is used where the initiating Promoter wishes to retain responsibility for certain aspects, but wishes to gain the benefit of private sector experience and funding for other parts.

Unlike PFI, payment can be made in the traditional way for capital investment required by staged instalments and/or by payment in a similar way to PFI, for availability of assets or services. The most high profile example of PPP in the UK is on the London Underground, where the private sector is responsible for the maintenance and upgrade of the fixed assets, whilst control of the rolling stock and provision of service remains with London Underground.

However, as with PFI, the contractual arrangements are complex, requiring similar governance and the creation of SPVs to support the Promoter.

Contracts for consultancy services

Traditionally consulting engineers in the UK and internationally were employed by Promoters under standard terms of engagement and fixed scales of fees stage-by-stage through a project. Other consultancy work for promoters or contractors was most often reimbursed at *cost plus* a fixed fee. Consultants are now increasingly invited to compete by price to provide design or other services, whether employed by promoters or contractors or as part of a joint venture/SPV for PFI/PPP arrangements.

Contract responsibilities

Responsibilities and duties of the Promoter

The Promoter's objectives, responsibilities, duties and liabilities should normally be stated in all contracts with consultants and contractors, including

- defining the functions that the project is to perform
- providing information and data held by the Promoter and required by the other parties
- obtaining the necessary legal authority to allow construction of the project
- acquiring the necessary land
- making payment.

The Promoter may of course arrange for some or all of these duties to be performed by a consultant or a contractor.

Responsibilities and duties of the Contractor

Contractor's responsibilities vary depending on the nature of the procurement route selected. In traditional, and design and build routes, contractors usually selected are those who, on account of their resources and experience, are able to undertake responsibility as Main Contractor for the construction of the whole of a project, although they may *sub-let* parts of the work to specialist or other sub-contractors.

Specialist contractors

In Management Contracting and Construction Management, contractors selected for the primary role to the Promoter may either be firms of main contractors with specialist divisions or be firms who have sought to build their business around MC and CM routes. The supply chain below the MC or CM is then usually drawn from contractors who confine their activities to selected classes of work and are referred to as specialist or *trade contractors*. This specialisation enables them to employ skilled staff and plant particularly suited to their work, but without carrying the same over-head requirements associated with main contractors. In some cases their designs and techniques are protected by patents.

International practice

International practice varies significantly across the various global regions, and for the purposes of this guide, it is difficult to give more than a flavour of the regional approaches.

In the eastward extension of the European Union, because of an historical tendency to use the public sector to both design and construct projects, the financial infrastructure lends itself to a PFI regime for major projects or programmes of work. The near European countries have favoured traditional contracting methods, often based around the FIDIC family of contracts, which are internationally recognised.

In the developing world, how a project is to be financed may affect how or which consultant and contractors can be employed. For these, the choice of is usually limited to those approved by a financing organisation, and the choice of contractors to those who can finance construction.

Major projects in the Middle East and Far East have historically been financed by traditional contracting, again using FIDIC.

A variety of systems is more typical on the American Continent, illustrating that there is no 'one size fits all'.

Sub-contracting

Sub-contracting is a term used to convey a relationship between the Main Contractor and others in the supply chain to deliver a project. A common principle is that a main contractor is responsible to the Promoter for the performance of the sub-contractors. Practice varies in whether a main contractor is free to decide the terms of sub-contracts, choose the sub-contractors, accept their work and decide when to pay them. It also varies in whether and when a Promoter may bypass a main contractor and take over a sub-contract.

Under most traditional forms of civil engineering contract, the Contractor is required to seek the approval of the Engineer before placing any sub-contracts. In design and build and CM/MC, different obligations may apply, but otherwise a common theme is that the Contractor or Trade Contractor shall remain liable for all the acts and defaults of sub-contractors.

On large Projects, containing many sub-projects within them, the sub-contracting ethos has been replaced by 'tiers' of Suppliers. Suppliers in tiers below the top level will have responsibilities to upward tiers and to the Promoter. This concept is known as an 'integrated supply chain'.

Other sub contracting may be predetermined by the Promoter or Project Manager. This is known as nomination and its prime uses are for long lead works or where a particular supplier's product is required, or if novel or risky work is required from a specialist sub-contractor. Care has to be taken in public sector contracting to ensure that EU procurement rules are not broken.

In some contracts for building work the architect or equivalent can nominate sub-contractors. In these arrangements the Contractor is instructed to obtain quotations from approved sub-contractors, and then accept the tender of the *nominated sub-contractor* and work with that sub-contractor as with any other. The Contractor usually has the right to decline to accept a nominated sub-contractor for a good reason.

The nomination of sub-contractors is not recommended, because it complicates relationships and divides responsibilities. Parallel contracts with the Promoter are the alternative arrangement, as is usual in industrial construction.

Internal contracts

A Promoter may choose to have work constructed by his own maintenance or construction department, known in the UK as *direct labour* or direct works, instead of employing contractors. If so, in all, but small organisations the design decisions and the consequent manufacturing, installation and construction work is usually the responsibility of different departments. To make their separate responsibilities clear, the order instructing work to be done may in effect be the equivalent of a contract that specifies the scope, standards and price of the work as if the departments were separate companies. Except that disputes between the departments would be managerial rather than legal problems, these internal 'contracts' can be similar to commercial agreements between organisations.

Many local government authorities in the UK have direct labour organisations (DLOs) which carry out at least small projects and maintenance work. Now compulsory competitive tendering (CCT) legislation requires their DLOs to compete for most work with independent contractors.

If a contractor promotes, as well as carries out, a project he may need to separate these two roles because different expertise and responsibilities are involved in deciding whether to proceed with the project and then how to do it. Separation of these responsibilities may also be required because others are participating in financing the project. For all such projects except small ones, an internal contract may therefore be appropriate to define responsibilities and liabilities.

Project execution plan and procedures

The results of the decisions taken on the contracting options should be the proposed strategy for a project. The Project Manager should agree it with all the managers who control resources and obtain the Promoter's acceptance of the strategy before proceeding with the project.

The strategy requires publicising to everyone working for the project, at least to the extent of stating the project objectives and scope. A small guidance note or a display specific to a project can achieve this. For larger or novel projects a 'project execution plan' or 'project implementation plan' and displays are required.

Whether or not a project needs a detailed separate plan or only a document that supplements established standards and procedures, every person who is to be responsible for planning and managing work for it should be told

- the Promoter's objectives
- the purpose of the project
- the performance criteria and constraints
- the quality and safety standards required
- the completion date, and any intermediate dates of importance to the Promoter
- the cost limits
- the priorities between time, quality and cost
- the risk and safety management policies, and any special requirements or constraints

- the organisation of the work – the work breakdown structure and the contract strategy
- the role and organisation of the project team and supporting resources
- the system for project communications, control and management.

The execution plan should be used to guide all the work that follows.

References

1. WEARNE S. H. *Principles of engineering organisation*, Thomas Telford, London, 2nd edition, 1993
2. PERRY P. *Risk Assessment: Questions and Answers*, Thomas Telford, London, 2003
3. INSTITUTION OF CIVIL ENGINEERS. *Managing health and safety in civil engineering*, Thomas Telford, London, 1995
4. JOYCE R. *The CDM Regulations 2007 Explained*, Thomas Telford, London, 2007
5. PERRY P. *Health and Safety: Questions and Answers*, Thomas Telford, London, 2003
6. PERRY P. *CDM: Questions and Answers*, Thomas Telford, London, 2nd edition, 2002
7. BADEN HELLARD R. *Total quality in construction projects*, Thomas Telford, London, 1993

5
Design

Design process

The aim of the designer should be to ensure that sufficient information is produced at each stage of the process to

- show that the project will achieve the Promoter's objectives
- obtain any necessary approvals and consents
- define in detail the next stages of the project, particularly the drawings and specifications needed for contracts, construction, testing and commissioning.

The successful design of a project demands not only expertise in technical detail, but also a wide understanding of engineering principles, construction methods, costing, safety, health, legal and environmental requirements. In the case of a river crossing, for example, should the solution be over, under, around or on? What will be the cost? How can it be built? How long will it take? How will it be maintained and what will the effects be on society and the environment? The resulting information should specify what is to be built, the standards and tests required, health and safety criteria, constraints on construction methods, and the life-long inspection and maintenance requirements.

As described in Chapter 2, design begins at the first investigation of ideas for possible projects with the project brief. It becomes progressively more defined through the investigation of schemes and alternatives. It usually proceeds in stages as illustrated in Fig. 7.

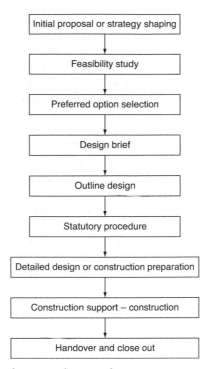

Fig. 7 Stages in development of project design

The design of a project can proceed through all these stages quickly if the project is small or similar to previous ones. Statutory Procedure may not always be required. Separate defined stages will be needed where the responsibility for design is to pass from one organisation to another.

Design brief

The Promoter's project brief will lead to feasibility studies, the designer must advise the Promoter or his Project Manager on the alternative possible schemes which might meet his needs. The

53

feasibility study will estimate the costs and the risks. As stated in Chapter 3 the result of the feasibility studies should be a preferred option which can be translated into a 'design brief'. This brief defines the scope, objectives, priorities and design criteria of the proposed project.

The brief should provide the guidance needed to undertake the preliminary design, but provide flexibility to develop the most cost effective solution. Changes to the brief should be made only if the information used in the feasibility studies has changed. During the preliminary design the construction techniques and any innovations should be considered. It is during this stage that the greatest savings can be made to the project costs. As shown in Fig. 4 in Chapter 3, the opportunity to make savings diminishes with time and later changes may require re-design, discarding of work done or delays to construction. Many considerations are likely to apply during the stages of the design process. Some of these are explored further below.

Design programme

The design brief should include a schedule of dates for delivering drawings, specifications, calculations and other information for cost estimating, contracts, ordering materials and, of course, construction. If the project is complex, uncertain or urgent, every stage may need detailed planning and coordination with site investigations, risk studies, modelling work, the timing of submissions for approval by the Promoter and planning authorities, environmental impact studies and the health and safety plan.

The design programme should consider when a contractor is to be appointed and their input to the design. It should give time for the team to visit the site to assess design and construction problems and if needed to arrange for surveying, site investigation, traffic counts, flow measurements and further studies such as the location of supplies of materials, use of local services, temporary traffic needs and so on.

The designer should advise the Promoter or his Project Manager of the time needed to prepare designs, contract drawings and a

54

specification, as lack of time for these is highly likely to lead to delays and additional expenditure. The Project Manager should advise the designer of any constraints including the time needed for consents and approvals by the Promoter. The responsibility for coordination would fall to the Construction Manager or Management Contractor in the event that this contracting option is adopted by the Promoter.

Quality-cost-time

Of all the considerations, arguably the most important is the relationship between the quality and performance of the completed project, its cost and the time taken for the work, within the requirements of health and safety. The need to consider these together is illustrated in Fig. 8.

Directions on balancing the cost, quality and speed of construction should be given in the design brief, as the achievement of one can conflict with the achievement of the others.

Health and safety

The design must comply with health and safety legislation. The main duties of the designer under the CDM Regulations are to

- Make sure that they are competent and adequately resourced to address the health and safety issues likely to be involved in the design.

Fig. 8 Relationship between quality, cost and time

- Check that Clients are aware of their CDM duties.
- Consider during design, avoiding foreseeable risks to those involved in the construction and future use of the works. In doing so, they should eliminate hazards (so far as is reasonably practicable, taking account of other design considerations) and reduce risk associated with those hazards which remain.
- Consider measures which will protect all workers if neither avoidance nor reduction to a safe level is possible.
- Ensure that drawings, specifications, operations and maintenance instructions, etc., include adequate information on health and safety.
- Provide adequate information about any significant risks associated with the design.
- Coordinate their works with that of others in order to improve the way in which risks are managed and controlled.

Aesthetics

The Promoter's brief to the designer should state the policy on the aesthetic aspects of the design. Aesthetics should be considered by the designer. For example, materials need not be visible in their base form. More attractive finishes can greatly improve the quality and durability of the finished product at little extra cost. Novel designs need to achieve a balance between whole-life cost and aesthetics.

Roles and organisation

The brief should also state who is to be responsible for the remaining design work. Roles in the design team need to be defined; in particular who will lead the team.[1] Specialist design contributions may be necessary. Multi-disciplinary team working can be effective provided that there is no confusion of roles or duplication of effort.

Under the CDM Regulations, the Promoter has a duty to appoint designers who have the competence and resources needed for their duties under the Regulations. The design team may include members of the Promoter's permanent staff, their Project Manager or external personnel, specialist firms or contractors employed for consultation and for design of sections of the project. In some cases the designer may be employed by the Contractor and not by the Promoter. The design of specialist work may be contracted to firms who later become nominated sub-contractors for the design and construction of parts of the project, but the designer should retain ultimate responsibility for the design of the project as a whole.

The design team should be located where the main issues need to be resolved, and therefore on a complex project some or all of the team may have to move to the project site at the start of construction.

Outline design

Outline designs giving more information than used for the feasibility study may be required for approval by the Promoter, the funders and statutory authorities. If so, the designer should provide the Project Manager with the information required for the Promoter and for obtaining planning and other approvals and consents.

Planning permission may be granted in one or two stages; first for an 'outline' submission which may be accepted subject to conditions, and secondly for a detailed submission. Time and resources are needed in order to achieve this.

The designer will also have to prepare, or cooperate in the preparation of, an Environmental Impact Analysis, with evidence that outside parties who may be affected by a project have been consulted and their interests considered. For this and other work which draws on specialist knowledge, the designer in cooperation with the Project Manager has to plan and manage interfaces with other professionals, the public and other third parties, in order to obtain information and agree a design which meets the Promoter's objectives and priorities. The Engineering Council provides engineers with a code of professional practice on environmental issues.[2] This includes guidance on the need to seek ways to change, improve and integrate designs, methods, operations, etc. to improve the environment.

Design development

Designs must be economic to construct. The availability of appropriate resources – materials, labour and plant – for construction is important to the economics of the design, particularly at geographically isolated sites.

The experience of contractors can be invaluable in achieving practical and effective cost and programme solutions. Construction expertise should be applied throughout the design process, rather than alternative designs being received from contractors with their *tenders* just before construction. Traditionally in the UK the experience of contractors was brought in only at the tender stage. Consultation with contractors during the design development is now preferred by many Promoters to ensure that the design is suitable for economic and safe construction. In some cases the Contractor is appointed early in the design process, in a process commonly known as Early Contractor Involvement (ECI), but if not the consultations need to be conducted fairly between contractors who may later be competing for a contract to construct the project.

The extent of design information needed for inviting tenders depends on the method for procuring construction. The Project Manager should therefore ensure that the type of contract to be used for construction is decided in good time for the designer to be able to produce the type and extent of the documents required for inviting tenders.

In the traditional procedure, the detailed design of a project should be completed before tenders for construction are invited. If contractors are being invited to be responsible for detailed design as well as construction, the Promoter's designer should assist in their pre-qualification, tender analysis and selection. Only a performance specification may be needed for inviting tenders, but the Promoter will usually employ a design team to check the detail. If the Contractor is appointed during the outline design to assist in obtaining the necessary statutory approvals, the designer will work with the Contractor to develop the design and construction price.

The Project Manager will then review the construction price before giving approval to proceed.

Some site or other information may be known only when construction is under way and in such cases, redesign or supplementary design work during the construction phase may be unavoidable. Typical reasons for this are

- excavation in ground which proves to be different from that inferred from site investigations
- structures to house equipment, the details of which are unknown at the design stage.

Whole-life requirements

The designer will normally need to design for economic performance during the whole life of a facility after its construction, through the stages shown in Fig. 9.

Use and maintenance

Design which uses standard materials and components makes replacements cheaper and easier to obtain. This can be important

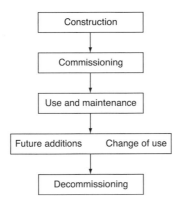

Fig. 9 Stages in the whole life of a facility

to the Promoter, not only to permit quick recovery in emergency situations, but also to obtain more components for a later expansion of the facility.

The costs of maintenance should be considered early so that the design can incorporate details which may be maintained economically. The designer will also have to consider if the facility can be maintained in a safe manner. If possible, experienced operating and maintenance staff should be consulted for their requirements for access to inspect, maintain and replace structures, materials or equipment throughout their useful life.

Future additions and change of use

Before proceeding to detailed design, the designer should suggest to the Project Manager how the design could include provisions which would facilitate future extension or changes. These provisions can produce a project of much greater value to the Promoter at little extra initial cost.

Decommissioning

Designs have finite lives and at some stage systems and structures have to be decommissioned and demolished. The designer can prepare for this by, for example, ensuring that structural frames are easily identifiable, specifying construction procedures that may be easily reversed, specifying materials that can be recycled or that generate low quantities of waste, and incorporating quality features into equipment design to safeguard against spillages of potential contaminants during operation.

Statutory approvals

The designer may need to present the design to the Promoter and representatives of statutory authorities, the public and others for their comments and approval. To get their approval the designer will need to understand the nature of objections to proposals and gain acceptance of the final design. On a large or novel project this may have to be done several times as the design is being developed.

The designer needs to develop good presentation skills and recognise the value of good presentation material. Time and training may be needed to develop these. The timing of presentations and consultations can be critical to the success of the project programme.

As an alternative to Planning Approvals some schemes require detailed Statutory Orders for the compulsory purchase of land or changes to existing statutory rights. A decision may be required by the Secretary of State and in some cases an Act of Parliament may be required. These procedures can take a long time and require a substantial amount of design, documentation and preparation. The scope and detail of the outline design may therefore be greatly increased. The designer should provide the Project Manager with the information required for the Promoter to present his case and obtain the necessary approvals.

Additional consents and approvals may be required from other statutory bodies such as utility companies or environmental bodies. Outline approvals and cost estimates sufficient to proceed to construction may be obtained based on the outline design, detailed approval may follow later. Again the designer should provide the Project Manager with the information required. In some cases the Contractor may obtain the detailed approval.

Detailed design

Detailed drawings and calculations are prepared for construction. Unless the project is urgent, the completed detailed design should also be the basis of a re-estimate of the cost of the project before being committed to a contract for construction. This re-estimate can be used to judge the prices that have come in from individual tenders.

Once the detailed design has started, changes should only be allowed if essential for the satisfactory completion of the project. A change control procedure should be applied so that the total potential costs and time effects of any proposed change may be determined.

The designer should be expert on specifying materials, recognising that ongoing research and development regularly yields new products. The engineering properties of materials should be understood and development work should be undertaken if needed to test new materials and methods.

Standard and locally available materials and components should be chosen wherever suitable, in order to reduce construction costs, make planning times more reliable and save training of construction employees. Innovations in construction methods and materials should therefore be considered early in design, in time to investigate their advantages to the Promoter's objectives. At the detailed stage it is wise to use proven technology, methods and materials for all time-critical aspects of a project.

Design methods

Calculation and analysis

Design is usually carried out in at least two main stages. For a building structure the first stage is to calculate the loads approximately with an analysis of the proposed design. When the analysis is checked and found satisfactory the loads are then calculated more accurately. Established standards and guides should be used wherever appropriate, to achieve reliable results economically.

The second design stage consists of proportioning the components of the structure according to the results of the calculations, together with an adjustment of the original calculations to any altered sizes of parts. The staged process is usual in the design of all types of project. Design iterations are often required to adjust parts of the design to detail from the designers of systems, plant and services.

Checking and assessment

All design should be checked, and the checking formally recorded on every drawing, specification and set of calculations. The assumptions and methodology employed in calculations should be subject to a review of principles. Assessment of all design by an independent party is recommended, and is statutorily required for larger structures.

Value engineering

Value engineering is an analytical technique for questioning whether the scope of a design and the quality of the proposed materials will achieve the project's objectives at minimum cost.[3] It can be used at every stage of design to check that the Promoter will get what he needs. It gives an opportunity for the designer to prove that design choices are economic and to identify where costs can be reduced.

Review and audit

Depending upon the project strategy on risks, the design process may be also monitored by independent review and audit. The application of a quality assurance policy may require audits during any stage of the design process. The designer needs to allow time for these audits and for any repetition of the work shown to be necessary.

Information technology

There are many computer-based tools to assist the designer to prepare and revise the design quickly and efficiently. Computer-aided draughting (CAD) is now universally used for studying alternative layouts, design coordination, checking clearances for construction, operations and maintenance, presenting the design, taking off quantities and revising detail. Other tools are available to test the critical load conditions and examine factors of safety.

The designer must understand how the computer packages employed operate, in order to be confident that the results are accurate. The designer must not lose the skill to apply engineering principles to determine solutions, and should seek to validate the operation of packages used and verify computed solutions.

Construction support

The designer is usually required to support site staff by

- clarifying the design and offering redesign, especially if unforeseen conditions are encountered
- checking that the assumptions made during the design are encountered on site
- checking and approving design carried out by the Contractor or sub-contractors, at least for *temporary works* which are the Contractor's responsibility
- specifying tests and assessing the results
- assisting in producing the as-built drawings.

The designer should therefore not only inspect the site thoroughly at the start of his work, but also visit it regularly during construction and be available until the project is complete and handed over.

References

1. RUTTER P. A. and MARTIN A. S. *Management of design offices*, Thomas Telford, London, 1994
2. ENGINEERING COUNCIL, LLOYD'S REGISTER and DEPARTMENT OF THE ENVIRONMENT. *Professional practice, engineers and the environment*, London, 1994
3. INSTITUTION OF CIVIL ENGINEERS. *Value engineering*, Thomas Telford, London, 1995

6

Contracts for construction

Contract formalities

In simple terms, a contract is an agreement that is enforceable by law. There are three basic formalities for the formation of a contract

- an offer and acceptance that is legally certain (for construction contracts this will usually require certainty as to the works to be carried out, the payment to be made (or at least the basis for assessing payment), the time period for carrying out the works and any other essential requirements)
- some form of 'consideration' (essentially something that constitutes value – normally a monetary sum, but a promise to perform another obligation may suffice)
- an intention to create legal relations (this is normally implied in a business setting and will therefore usually be a given for construction contracts).

So long as the above requirements are met, it is perfectly possible to have an oral contract. Furthermore, a contract may be implied where, although a formal contract has not been drawn up and signed by both parties, there is sufficient certainty as to what obligations each party is to perform and one party provides 'consideration' by, for example, starting the relevant performance.

Accordingly, care needs to be taken in issuing letters asking contractors or construction professionals to carry out work if it is not intended to create a binding contract at the time the letter is sent out. Furthermore, it is obviously important that contracts set out all the relevant terms that are agreed so that both parties are clear as to their respective obligations. Care is therefore needed to draw up contracts that incorporate all the relevant terms and letters of intent and similar less formal arrangements need to be carefully drafted to avoid unintended consequences of creating binding legal obligations going beyond those contemplated at the time the relevant letter is issued.

Contract contents

Contracts should specify the scope, location, quality and type of work to be carried out, any time period within which the relevant work is to be carried out and the relevant payment terms. With regard to the latter, care now needs to be taken to ensure that all contracts

(i) provide an adequate mechanism for determining what payments become due under the contract, and when

(ii) provide for a final date for payment in relation to any sum which becomes due in accordance with Section 110 of the Housing Grants, Construction and Regeneration Act 1996, failing which, the monthly payment arrangements will be implied under the Scheme for Construction Contracts.[1]

Contracts will usually seek to deal with the occurrence of foreseeable events beyond the control of the Contractor and/or the Promoter and will usually list risks that may occur during the contract period that will entitle the Contractor to seek additional payment and/or extensions of time to any date for completion.

Contracts should also deal with

- who is responsible for design, construction and supporting work
- how risks are shared between the Promoter and Contractor
- entitlement and any formalities relating, to the use of sub-contractors
- programmes of work and dates for completion together with provisions for extensions of time if liquidated damages are to be payable for delayed completion

- insurance arrangements
- terms of payment (see above)
- variations to the works to be carried out
- grounds for termination of the contract itself and possibly for the termination of the employment of the Contractor (in which case, the contract itself will continue with *The Employer* being entitled to appoint a replacement contractor to complete the works and contra-charge the original contractor for any additional costs)
- the settlement of disputes.

Standard forms of construction contract

Standard or model forms of construction contract have been published by a variety of institutions since the Second World War. Historically, there has been a distinction between the standard forms of contract issued for different types of work. The Joint Contracts Tribunal (JCT) have published contracts primarily intended for use in connection with building works whilst the Institution of Civil Engineers (ICE) have produced standard forms of contract primarily intended for engineering works. While many of the terms within the different suites of contract have been broadly similar, there have been traditional differences. For example, under JCT contracts, ground conditions are usually a risk of the Contractor (based on the assumption that contractors are likely to have local knowledge of ground conditions and/or that it will be possible in most cases to obtain ground investigation reports in relation to the area of land to be covered by the proposed building). In contrast, under ICE contracts, ground conditions have traditionally been an employer risk and give a contractor entitlement to compensation (based on the premise that in the case of civil engineering projects such as the construction of a road, it is simply not practical to obtain ground investigation reports for a road of even moderate length).

In more recent years, there has been a blurring of the traditional distinction between the forms of contract for building and

67

engineering works with contracts such as the NEC suite and the JCT-Constructing Excellence Contract being applicable to both disciplines. Indeed the NEC suite of contracts deliberately changed the title of the main form of contract from the *New Engineering Contract* to the *NEC Engineering and Construction Contract* to try to avoid any perception that it was a contract for engineering projects only.

All standard forms of contract have been drawn up after very careful consultation and consideration to ensure that their terms are broadly fair and are mutually consistent.

Notwithstanding this, there has been a tendency in the construction industry for individual Promoters to produce their own bespoke amendments to whichever standard form of contract they are using. Care needs to be taken that the amended documents remain clearly drafted and the provisions mutually consistent and, from a practical point of view, Promoters need to be aware of the potential effect of risk pricing if contract terms are amended in their favour.

Following the reports of Sir Michael Latham and Sir John Egan (*Constructing the Team* (1994) and *Rethinking Construction* (1998) respectively) there has been a movement towards the development of more collaborative forms of contract, specifically the NEC suite of contracts, the ACA Project Partnering Contract (PPC 2000), and the recently published JCT-Constructing Excellence Contract. All these contracts seek to establish, in addition to the usual construction obligations, an obligation on the parties to act collaboratively and consider the interface between members of a project team with each other, rather than concentrating solely on individual contracts with a Promoter with the risk of a contractually blinkered view of the role of other parties.

Care is required to select a form of contract which is appropriate to the proposed procurement approach and requirements. If a Promoter is seeking to set up a collaborative approach, one of the collaborative forms of contract is best used. On the other hand, if a Promoter does not want to manage a design team, a design and build approach may be more appropriate (albeit that this may follow a novation of the designers to a design and build contractor once the outline design has been developed to the satisfaction of the Promoter).

With the NEC, PPC and JCT – Constructing Excellence Contracts it is possible to use the same contract documents (completed in different ways) to fit with different procurement approaches whereas with some of the more traditional contracts there may be different forms for different procurement approaches – for example the separate JCT and ICE forms for design and build or for consultant led design.

Different publishing bodies for standard forms of construction contract

The Institution of Civil Engineers

For many years, the ICE has published the *ICE Conditions of Contract*. This is presently in its 7th Edition (published in 2003). It is based on a *remeasurement* approach where an estimated cost is provided (the 'Tender Total') by the Contractor and agreed by the Promoter, but thereafter, all works are re-measured to assess the quantities of works actually required and the 'Contract Price' finally due to the Contractor. The contract is now part of a suite comprising contracts for

- Minor Works
- Ground Investigations
- Design and Construction.

In 1992, the ICE (via its NEC Working Group) first published the *New Engineering Contract* which was designed around project management procedures of cost and programme management. At the time of its issue it was considered to be radical and was very different to the other standard forms of engineering contract, not least because of its use of language – it was, and remains, written almost exclusively in the present tense, even when indicating matters that are intended to take place in the future. The Latham Report in 1994 considered the contract to be the best of the then standard forms of construction contract and expressed the hope that the large number of standard forms of construction contract would reduce. The ICE issued a new edition entitled the *NEC Engineering and Construction Contract* in 1995 which addressed the two areas in which the Latham Report considered it to be slightly lacking and, with the latest edition 3 ('NEC3') published in 2005, it has now become one of the most used forms of contract in the construction and engineering industry.

The Joint Contracts Tribunal

The Joint Contracts Tribunal or JCT, as it more commonly known, has been producing standard forms of contract since shortly after the Second World War. The JCT is composed of a number of representative organisations, representing Promoters, contractors, engineers, architects and cost consultants. The main form has been the *Standard Form of Contract* which has been published in a number of different editions and codified as its *Standard Building Contract* in 2005. The 2005 version comprises 4 different versions:

- With Quantities
- With Approximate Quantities
- Without Quantities
- Design and Build.

The *Standard Building Contract* is also part of a wider suite of contracts comprising

- Minor Works
- Intermediate Contract
- Major Project Contract
- Management Building Contract
- Construction Management Building Contract.

In 2006, the JCT published the *Constructing Excellence Contract*. In perhaps a similar way to the difference between the ICE traditional forms of contract and the NEC suite of contracts, the *Constructing Excellence Contract* is markedly shorter and easier to read than the traditional forms of JCT contract.

The Office of Government Commerce

Until 2005, the Office of Government Commerce (OGC) had taken over responsibility for the publication of the long running Government Contracts – the 'GC/Works' suite of construction contracts. In 2005 it announced that it would not be updating the GC/Works suite and that the contracts would no longer be available for purchase. Notwithstanding this move, a number of Promoters continue to use the GC/Works forms of contract which are intended to deal with both building and civil engineering projects, with different versions for a variety of different types of project and procurement arrangement, including minor works and design and build forms.

The OGC allowed an endorsement to the latest NEC (NEC3) suite of contracts that the suite complies with the principles of the OGC's *'Achieving Excellence in Construction'* guidance. This endorsement was intended to direct public sector Promoters to the use of the NEC3 suite in place of the GC/Works suite of contracts. However, as at spring 2008, the OGC is presently undertaking a further independent review of the PPC 2000 and JCT – Constructing Excellence contracts to see if this endorsement should now be given to all or any of these three contracts.

Fédération Internationale des Ingénieurs-Conseils (FIDIC)

FIDIC has been publishing engineering and building contracts primarily intended for use in relation to international projects. The present suite of contracts comprises

- the new Red Book, Conditions of Contract for building and engineering works designed by the Employer
- the new Yellow Book, Conditions of Contract for plant and design-build for electrical and mechanical plant and for building and engineering works designed by the Contractor
- the new Silver Book, Conditions of Contract for EPC and turnkey projects
- the new White Book, Client/Consultant Model Services Agreement.

71

The Conditions of Contract are, compared to most forms of domestic construction contracts used in the UK, very onerous in passing the majority of the risk of project delivery to the Contractor. This is an accepted bias in the world of international projects where the Promoter often expects to receive the outcome contracted for without having to get involved in the detailed management of a project.

Mechanical and electrical and process plant contracts

There are a number of bodies who publish standard forms of contract for use in relation to projects involving the installation of process plant and related equipment. Projects will usually involve building works but these are often of far less value than, and ancillary to, the plant being installed in the relevant buildings. As a result, greater attention is focused in the contract documents on inspections of the relevant plant and its passing of tests prior to delivery and following installation to ensure the relevant plant performs as intended.

Two of the best known contracts are

- the 'MF/1' Model form of General Conditions of Contract for use with home or overseas contracts for the supply of electrical, electronic or mechanical plant – jointly published by the Institution of Electrical Engineers (now the Institution of Engineering Technology) and the Institution of Mechanical Engineers
- the suite of contracts (the 'Green' and 'Red' Book Conditions of Contracts for Process Plant for 'Reimbursable' and 'Lump Sum' contracts respectively) published by the Institution of Chemical Engineers.

Standard form of contract for PFI and Public Private Partnership projects

Over recent years there has been a move towards the greater standardisation of contracts under the Private Finance Initiative. The Treasury has published standard contractual terms for PFI projects (SoPC4) which sets out the standard approach to risk allocation between public and private sectors and includes mandatory principles and drafting for certain key contractual clauses.

While SoPC4 is only mandatory for PFI contracts, many of the contractual provisions and commercial positions set out in it are equally applicable to other forms of PPP and the Office of Government Commerce are keen that procuring authorities consider SoPC4 when negotiating contracts for alternative PPP solutions, particularly where private finance is used to fund all or an element of the capital requirements for the project.

The range of PPP contractual arrangements has developed over the last ten years, though all tend to involve the private sector accepting the majority of the risk of project delivery with arrangements for the provision of services and the sharing of risks beyond the construction period. The contractual structures are often referred to by different acronyms, such as BOO (Build Own and Operate), BOOT (Build Own Operate and Transfer), DBOM (Design Build Operate and Maintain), DBFM (Design Build Finance and Maintain) and DBFO (Design Build Finance and Operate). In different sectors, organisations such as 4Ps (Public Private Partnerships Programme) have developed standard forms of documentation for particular markets – for example, there is a standard form of DBOM contract for leisure projects.

Underlying considerations for successful contracts

While all construction contracts need to be clear and relevant to the works being carried out, Promoters and their advisers need to have in mind the underlying considerations of

- value/functionality
- risk
- cost.

Increasing value/functionality

The 'value' or functionality that a construction project will bring to a Promoter requires not simply consideration of the capital costs of

73

construction, but also of the through life cost of owning and operating a building, structure or item of plant as well as consideration of how the new building/structure adds to the achievement of the Promoter's business. Consider the example of the call centre operator who achieved a quantum leap in the performance of its business not by simply improving the efficiency of the construction of its buildings, but by re-engineering the layout of its buildings so that they were more enjoyable places to work in. Although the capital cost of the buildings increased by a small fraction, the resulting reduction in staff churn reduced overall business costs by a much larger fraction.

Historically, the construction industry has concentrated almost exclusively on the initial cost of projects and (as is explained below) more particularly on the *price* of construction projects.

Risk and cost/price

The issues of risk and price have traditionally not been properly separated. Historically, Promoters have amended contracts to pass greater risk to Contractors without considering the effect this may have on the Contractor's pricing of the works to be carried out. Increasingly, Promoters are coming to recognise that passing risk usually has a cost consequence. Sometimes this is recognised and accepted as a worthwhile trade-off for not having to manage a particular risk. However, the level of risk pricing is often not clearly understood with the result that sometimes Promoters do not realise the hidden additional costs they are paying for passing risks, which they might themselves be able to manage effectively at less cost.

For example, under building (as distinct from civil engineering) contracts, the risk of ground conditions is often considered to be an appropriate risk for contractors to manage. If a contractor is not entitled under the contract terms to compensation for the occurrence of adverse unforeseen ground conditions, it is likely that they will include in their tender price some risk contingency to cover the potential of this risk occurring. If the amount of this risk contingency is disclosed, in many situations Promoters might prefer

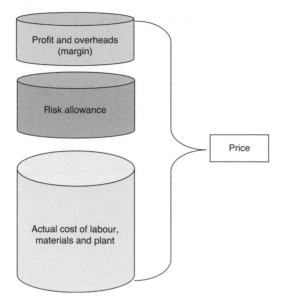

Fig. 10 Elements of cost in the construction industry

to do something about reducing the amount of the contingency or removing it entirely by taking on themselves responsibility for unforeseen ground conditions. Accordingly, it is suggested that a better, more cost effective, approach may be to try to identify the major risks affecting a project and do something to mitigate them before asking a contractor to price for them. For example, in the above example, obtaining ground investigation reports in advance of entering into the construction contract will provide both Contractor and Promoter with a better understanding of the potential risk and the level of contingency that may be appropriate to deal with this risk.

The construction industry is used to dealing in 'prices' which will in fact include separate elements, as illustrated in Fig. 10, in respect of

- profit and overheads

- a risk allowance
- the actual cost of labour, plant and materials.

Often, the distinction between the three elements is blurred, and as a result, rather than starting from an understanding of the actual cost of the works, and building up an understanding of the risk contingency appropriate for a particular project (with a profit margin on top), prices are often calculated on the basis of composite prices based from previous similar projects. There is a great deal of data within the construction industry regarding prices for different forms of activity but far less information regarding actual costs (both initial costs and through life costs).

In an attempt to encourage the discipline of distinguishing between the above three elements, some modern forms of contract seek to adopt payment by reference to 'actual cost' rather than *valuation* by reference to rates and prices for different types of work.

Letting construction contracts

The choice of the form of contract to be used for a particular project is usually made by the Promoter (with or without input from its advisers).

Traditionally, the majority of construction contracts have been tendered, with a Contractor providing relevant information (usually a price, a programme and any necessary design proposals) on the basis of accepting a specified form of contract stated in the tender.

However, with some Promoters who are seeking to set up colla-borative arrangements, a Contractor may be given the opportunity to comment on any terms in the proposed form of contract that create any particular risks that may not have been considered by the Promoter although it is very unusual for a Contractor to be asked to put forward his own choice of contract in preference to the Promoter's choice. Negotiation can have the advantage of allowing greater consideration of underlying risks and ways of removing or mitigating them before asking contractors to price for work. Even in tender situations, it may be possible to request different prices for different allocations of risk which may at least provide a greater understanding of potential risk pricing.

Ancillary issues

Collateral warranties

Collateral warranties are intended to create contractual obligations between parties who would not otherwise be in direct contract. In the case of a Promoter engaging a Main Contractor, the Promoter will obviously be in direct contact with the Contractor, but will not be in a direct contractual relationship with the Contractor's sub-contractors. However, by entering into a collateral warranty with the sub-contractors, the Promoter can obtain a contractual remedy directly against the sub-contractors in the event of breach of the collateral warranty.

Collateral warranties traditionally contain a statement or 'warranty' that the party giving them has performed the terms of their relevant contract (be it professional appointment or construction contract). There is usually also an obligation to maintain a minimum level of professional indemnity insurance to provide comfort to the party receiving the warranty that the person giving the warranty has insurance backing for any claim made under the warranty.

Following the Contracts (Rights of Third Parties) Act 1999, it is possible to create rights in favour of third parties to a contract without the necessity to enter into separate collateral warranty arrangements. Examples of such arrangements are set out in a number of the latest versions of JCT contracts.

Security for performance

Performance bonds

Performance bonds may be requested from contractors or sub-contractors to provide security for the performance of their respective contractual obligations. A performance bond will usually be provided by a bank or insurance company and will usually be limited to a percentage of the value of the works being carried out under the relevant contract. There is a cost in providing a performance bond which is usually passed to the recipient as part of the works costs. Nowadays, performance bonds will not be payable 'on demand', instead, the recipient of the bond will need to demonstrate that loss has been suffered as a result of breach of the relevant contract. Sometimes, depending on the wording of the bond, this may require proof by way of an adjudication or court order that a breach of contract has occurred and that a specific sum of compensation is now due.

Parent Company Guarantees

As an alternative to performance bonds, Parent Company Guarantees may be requested by Promoters from Contractors and/or subcontractors where they are subsidiaries to a larger parent company. Liability under Parent Company Guarantees may be expressly limited but often is not.

Reference

1. The Scheme for Construction Contracts (England and Wales) Regulations 1998 (Statutory Instrument no 649 of 1998)

7

Planning and control of construction

Responsibilities

This chapter describes the main tasks of planning and controlling the construction stage of a project. Who is responsible for each task depends upon the contract arrangements. Figure 11 indicates how these tasks are shared between the Promoter, consulting engineer and Contractor in a traditional contract.

Reference to what is considered a traditional contract is made in Chapters 4 and 6. There are alternative forms of contract which allow for the interaction of the contracted parties and their representatives to varying degrees.

The tasks shown in Fig. 11 are more the responsibility of a contractor in design and construct contracts. They are completely a contractor's responsibility if they are promoting the project and undertaking its financing, design, construction and commissioning. In a contract between a promoter and a joint venture of contractors, the contractors are usually individually and jointly liable as if they were one contractor.

Initiation of construction

Starting date

In most contracts the Promoter or his representative under the contract notifies the contractor in writing of the date for starting the project. The Contractor is then responsible for proceeding with

Civil engineering procedure

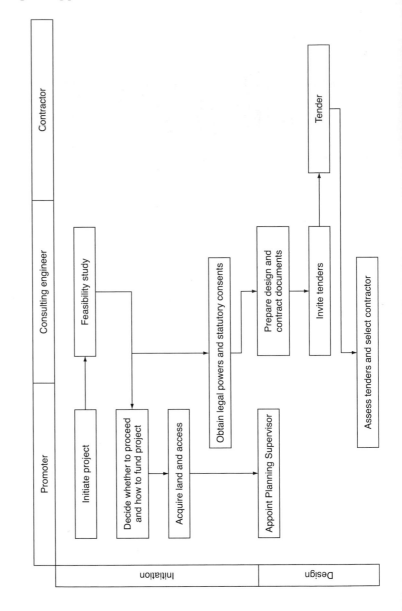

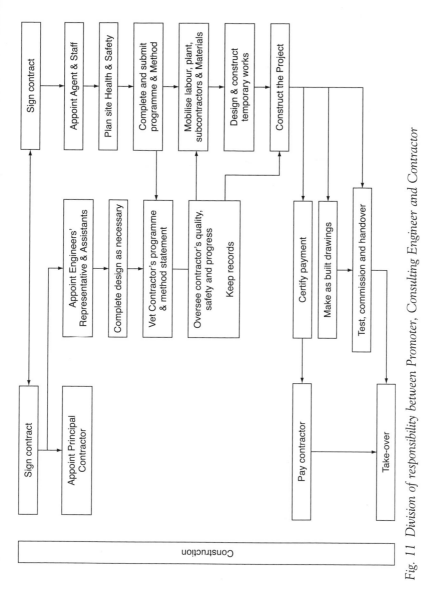

Fig. 11 Division of responsibility between Promoter, Consulting Engineer and Contractor

due diligence and completing the project in accordance with the contract.

Under the UK CDM Regulations the Promoter has to appoint a Principal Contractor and has the legal obligation not to permit construction to start until the Principal Contractor has prepared a satisfactory Construction Health and Safety Plan.

Statutory notifications

Statutory planning approval for a project is usually obtained by or on behalf of the Promoter. Promoters and contractors must meet legal requirements to notify new activity to statutory bodies such as the Health and Safety Executive, the local Environmental Health Officer and the Fire, Police and Ambulance Services.

The Promoter's representative

In the traditional type of contract for the construction of a project the Promoter must name a person to be the Engineer (ICE Conditions of Contract) or Project Manager (NEC). The person appointed then has duties and powers to administer the contract.

In the ICE and the NEC conditions, the Representative's decisions in administering a contract must be impartial as between Promoter and Contractor, and must be based on the terms of the contract. Any restrictions imposed by the Promoter on the Representative's authority to exercise his powers stated in a contract must be notified to prospective contractors when inviting tenders, or negotiated between Promoter and Contractor during the contract as a change to its terms.

When supervising the work to see that the Contractor has executed his work in accordance with the contract, the Representative is also the agent of the Promoter, thereby acting in a dual capacity. The Representative normally reports to the Promoter monthly, including a review of the progress of the project, major decisions and impending important events.

In some forms of contract, the contractor will deal directly with the Promoter, who will appoint members of their team to liaise with the contractor and will ensure that the Contractor is fulfilling his obligations.

Programme and methods of construction

In the ICE conditions the Engineer, as the Promoter's Representative is required to approve the programme of construction and consent to the methods which the Contractor proposes to use. The Contractor must provide these under the contract if requested by the Engineer. This is called the Contract Programme against which the progress of the project is monitored.

Similarly in other forms of contract, the client will be provided with a programme of works.

Supervision

Under the ICE Conditions, the Engineer usually appoints a representative to be on site (the *Engineer's Representative*) to watch and supervise the construction and completion of the project. Under the NEC conditions, this is the *Supervisor*. This Representative must look ahead and discuss future parts of the project with the Contractor's manager in charge at the site to ensure that they are planned to achieve the approved programme. (*Agent* was the traditional title in the UK for the Contractor's Representative and manager in charge on site. However, the title Project Manager or Project Director is now commonly used by contractors. These titles are not used in this guide to avoid confusion with the Promoter's Project Manager).

Changes and variations

Figure 4 in Chapter 3 indicated that changes to design during construction can be expensive, because of the direct costs of repeating work and scrapping materials and the indirect costs of

83

disrupting economic working. In the traditional procedure in the UK the Promoter should not propose a change to the Contractor directly. It should be discussed with the Promoter's Representative and its benefits compared to cost, on the basis of its value in achieving the objectives of the project. If the Promoter then wishes to make the change, the Representative may be able to instruct the Contractor to proceed with it, depending on the terms of the contract, or may have to negotiate it with the Contractor.

It is considered good practice that all variations and changes should be recorded and agreed as works progress. The effect of these changes may be considered detrimental to the progress of the works and commonly form the basis of a claim.

In the ICE conditions of contract the Engineer can instruct the Contractor to vary the Works if they think it necessary or desirable. Depending on the terms of the contract, variations may include instructions to add or omit work or change the contract programme. The Engineer should then instruct the Contractor by means of a formal Variation Order (VO). He should advise the Promoter on variations found necessary or desirable and inform the Promoter of their effects on the programme and the cost of the project.

VOs should specify the varied work in detail. The prices to be paid for new or additional work and other effects of a variation should be stated in a VO, if the Representative has the power in the contract to do so, and if not, negotiated with the Contractor. VOs may be prepared by the Representative's staff, but may be signed by them only if empowered to do so. In traditional *admeasurement* contracts, changes from the quantities of work stated in the *bill of quantities* which result from drawings issued by the Representative during the contract are not 'variations'.

Changes proposed by the Contractor

A change proposed by the Contractor should be considered by the Representative and a recommendation as to whether or not to accept it made to the Promoter. If the proposal is accepted, the

Representative makes it a variation ordered under the contract. In some contracts the Contractor has a duty to propose variations which he considers may be necessary for the satisfactory completion of the project.

Changes negotiated between Promoter and Contractor

The parties to a contract can agree changes to it at any time, separately from powers that the Representative (Engineer in the ICE conditions) may have to order variations. This is necessary for a change which, for instance, the Representative is not authorised by the contract to order as a variation.

Completion certificates

When, in his opinion, the project has been substantially completed and passed the relevant tests, the Representative under the contract is required to issue a certificate to that effect. The *defects correction period* then normally begins.

A completion certificate may also be issued for a completed part of the project. If requested by the Contractor, the Representative must issue a completion certificate for a substantial part of the project if it is completed to the Representative's satisfaction and is being used or occupied by the Promoter or anyone acting on his behalf or under his authority.

The project must be handed over to the Promoter at the end of the defects correction period in the condition required by the contract, and the Contractor must complete any outstanding work and also make good any defects during the defects correction period or immediately thereafter.

If any defects for which the contractor is responsible are not corrected in this period, the Promoter is entitled to withhold from the balance of the *retention money* due to the Contractor the estimated cost of such work until the Contractor has completed it. Failing this the Promoter may arrange for it to be completed by others at the Contractor's expense.

Responsibilities of the Contractor

The responsibilities of a contractor depend upon the terms of a contract and the relevant law. The following notes are written on the basis that one main contractor is 'the Contractor' in control of the site, but they also apply to each of several contractors if they are working on a project in parallel.

Implementation of the works

The Contractor is responsible for constructing and maintaining the project in accordance with the contract drawings, specification and other documents and also further information and instructions issued in accordance with the contract.

The Contractor should be as free as possible to plan and execute the works in the way he wishes within the terms of his contract. So should sub-contractors. Any requirements for part of a project to be finished before the rest and all limits to the Contractor's freedom should therefore have been stated in the tender documents.

Sub-contractors may have varying status within the contract. They can be directly employed by the Contractor or may be nominated by the Promoter or their representative under the contract. These will be under the direct control of the Main Contractor. Other contractors may be appointed by the Client with their activities coordinated by the one Principal Contractor.

Health and safety

The Planning Supervisor appointed by the Promoter earlier in the project and the Principal Contractor should initiate meetings before construction starts to review the proposed methods of construction to identify hazards, assess the risks and minimise their dangers. As noted above, the CDM Regulations, 2007 will make excellent further reading with regard to the Health and Safety responsibilities of all parties.

Representatives of all parties on a site should attend regular meetings to consider health and safety needs, plan preventative measures, arrange training and hear reports and recommendations on any accidents and near misses. Feedback should also be received from regular safety inspections and tours, with particular reference to any developing patterns.

Successive contractors may need to be named as the Principal Contractor if possession of the site passes from one to another. On a multi-contract site the Promoter has to name one contractor as Principal Contractor to supervise health and safety over the whole site, as illustrated in Fig. 5 in Chapter 4.

Insurance

In most contracts the Contractor must insure the project until it is handed over to the Promoter.[1] The cover is usually in the joint names of the Promoter and the Contractor. It is however increasingly common for the insurance of the works to be held by the Promoter, when more advantageous terms can be negotiated.

Construction planning and control

Before the start of construction a scheme of work should be planned by the Contractor's senior staff who will be directly responsible for its execution. Decisions should be made on construction methods, site layout, temporary works, plant and the like, and requirements for labour, materials and transport.[2] The layout of temporary works areas, buildings, offices, accommodation, stores, workshops and temporary roads and railways needs attention, because the location of these features in relation to the project can greatly affect the convenience and economy of future construction and administration.

Outline programmes prepared by the contractor for tendering for a project are not likely to be based upon detailed study of the use of resources for the actual execution of the work. Detailed planning is normally needed at the start of construction in order to decide how to use labour, plant, materials, finance and subcontractors economically and safely.

As noted earlier, one of the first contractual duties of the Contractor is to submit a programme for approval by the Promoter's Representative under the contract. This programme should show the periods for all sections of the project so that the Representative can be satisfied that everything can be completed by the date specified in the contract. The Contractor is also required to submit a general description of his proposed method of work. If required by the Representative, they must be amended by the Contractor and resubmitted at the earliest possible date.

Note that 'programme' as used here means a diagram or table showing when work is to be done, as distinct from a computer software 'program'. 'Schedule' is an alternative name for a programme of work used particularly in the USA.

The programme should show the Engineer when any further information, drawings or instructions will be required, and the dates when various sections of the project will be completed and ready for use or for the installation of equipment by other contractors. All staff on site should review the programme and progress regularly to look ahead to check that the project will be completed to the date specified in the contract.

Methods of programming

The most widely used forms of programme are bar charts and network diagrams. Bar charts (such as Fig. 2 in Chapter 1) can show programmes in a form that can be easily read and then used to compare progress with planned dates. Network diagrams show the sequence and interdependence of activities and indicate the effects of delays. Networks may be drawn as an arrow diagram or a precedence network. Either can be used to calculate the critical path of activities which determine the total time to complete all the work. A very simple example is shown in Fig. 12.

Modern software is particularly sophisticated. This allows for programmes and charts to be shown in alternative modes such as

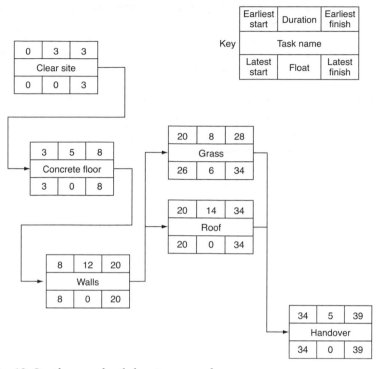

Fig. 12 Simple example of planning network

bar charts, networks and time-location charts. The growing power of computers is increasing the number of options.

It should also be noted that plans can now be resourced to a high degree of detail and that many uses can now be made of this data. For example, the fully resourced contract programme is increasing becoming a tool for use by the commercial team.

Many computer-based packages are available for displaying the network and the critical path of activities for large and complex projects. They can be used to analyse the use of resources, review progress and forecast the effects of changes in the timing of work or the use of resources. The choice of a computer package should be made after considering how far to integrate time and cost data for planning and control and also the need to feed back the resulting data into a database for planning and costing future contracts.

Programmes

Depending upon the duration of the contract, consideration should be given to the type of programme or plan and the amount of detail contained therein. For example, a project of approximately two years duration would benefit from long-, medium- and short-term plans.

The long-term or over-all contract programme will reflect the project requirements and will be used to pick up major items such as sub-contractors, strategic and long call-off procurement items. This plan will cover the whole duration of the project and will usually be the one that the Contractor will supply to the Promoter's Representative or Project Manager for agreement. The form of this programme can vary and may be a bar chart or a time/location chart, if the project under consideration is a linear, such as highway construction.

A medium-term plan, which would traditionally be called the 13 week, 3 month or 100 day plan, is particularly useful for ensuring that materials are called-off in time and notices are given to sub-contractors on a timely basis. This plan will support the overall, long-term plan and will be used to progress the works accordingly.

A weekly plan, which could be a 2–4 weekly rolling plan, is used to ensure that every last detail is in place for current construction. It will include details such as supervisor's names, items of plant and detailed materials quantities.

All of the above are supported by method statements and other project specific documents such as the Construction Phase Health and Safety Plan.

Since most planning packages are computer based, all three of the above programmes can be taken as a section or slice of an over-all plan. This has the advantage of being a single plan which can be updated as required.

Detailed programmes

As noted above, every section of the project should have its detailed programme to ensure that the work is planned and methods and needs for materials are agreed in good time. Programmes should be limited in size to avoid confusion and to assist communication and understanding. Most computer-based planning packages can produce critical path diagrams and other results of data for 10,000 or so activities. In order to be able to understand the results it is usual to show no more than 50 activities in one programme, and to display the results in the form of a bar chart as illustrated earlier in Fig. 2 in Chapter 1. For all projects except small ones, the use of a hierarchy of programmes is therefore recommended, with one activity in a high level programme summarising detail shown in a lower level programme.

Resource levelling

The Contractor will normally subject his draft programmes to a 'resource levelling' study to minimise the costs of peaks and gaps of the use of plant and labour. Computer-based systems are particularly useful for resource scheduling.

Labour planning

Contractors' estimates of the costs of work prepared for tendering are usually based on labour productivity for each type of work. These and the planned rate of work provide a basis for estimating the total labour. A chart of labour requirements can then be produced, showing by categories (skilled trades and labourers) the total numbers expected to be required at any particular time.

Economy in the use of plant and labour is achieved by planning to use it continuously and to maximum capacity. Planning and control should be in sufficient detail to see that expensive plant will not be idle for the want of adequate manpower or of transport of materials to or from the plant.

Plant planning

The construction plant that will be required and the periods during which it will be employed must be determined as early as possible, in conjunction with decisions on the detailed sequence of work and site layout. The time to obtain plant can be critical, depending on what plant is available from other contracts, what new plant should be purchased and what hired. For construction in many developing countries the need for supporting training and maintenance facilities has to be considered in deciding the choice of plant.

Programming the use of plant can be based on statistical data on the potential output for each machine and assessing the risks of inter-ference and changes. As the work proceeds, data on actual output achieved with each major machine or set of machines should be analysed. The causes of poor output should be found, remedied, and a new forecast made of when the work will be completed. Records of output achieved should also form the basis of statistics for planning future work.

Another type of planning document is the time/chainage or time/location chart. This is used to ensure that the correct sequence and logistics are in place and are useful for physically long projects, such as motorways and smaller jobs which are logistically challenging.

Materials planning

Materials can typically account for anywhere between 30% and 60% of the cost of construction. Their purchasing and use therefore need to be planned and controlled accurately; starting with detailed programmes which should enable buyers to draw up a schedule of materials required and ensure they are delivered on time. Materials planning is now often carried out with the assistance of specialist procurement departments.

Sub-contracting planning

Sub-contractors should be appointed well in advance and their programmes obtained to show that they can perform their work properly and safely in the time allowed.

Integrated planning

For larger projects, with complex contractor interfaces, an integrated programme may have to be assembled to ensure that all interfaces are managed successfully. This will not change the responsibility of the Contractor for their programme.

Modifications to programme

Approval of a programme does not mean that it cannot be changed. A good programme is flexible enough to permit modifications to meet the more probable risks. Experience shows that a programme which allows for contingencies enables those in charge of the work to see what the effect of adverse events will be on subsequent work and adjust their plans accordingly. The working programme should therefore be updated regularly. Revisions will also be required if the work is varied, acceleration is required, or extra time is given for any reason. The Contractor should then submit a modified programme for the approval by the Promoter's Representative under the contract.

Design of temporary works

The Contractor is usually responsible for the design of the temporary works he proposes to use, but these are of course subject to legal requirements for health and safety and specific regulations for independent checking of major falsework and other temporary structures.

In the ICE conditions of contract, the Contractor is required to submit drawings and design calculations for temporary works to the Engineer. The Engineer should scrutinise these with care. This in no way relieves the Contractor of his responsibility for the design and construction of the temporary works, but it should reduce the risk of mistakes and helps the Engineer to discharge his responsibility to the Promoter to see that the work is done satisfactorily and safely.

Quality

Contracts usually specify in detail the quality of materials and workmanship required, and the tests required to prove compliance. When planning the construction work the Contractor should make sure that the proposed methods and plant can produce work to the quality specified. During the work the Contractor's senior management have to ensure that quality is achieved despite pressures to achieve progress in bad weather or other adverse conditions. Most contractors have quality assurance systems which are designed to ensure this for all their work, and will in many cases, produce a site specific Quality Plan, which may be included in the overall Health, Safety Quality and Environment Management Plan.

Setting out

The Contractor is usually responsible for setting out the works. The Promoter's Representative or assistants normally check the setting out in order to minimise the risk of errors and consequent delays to progress. If the setting out is particularly complex, a specialist sub-contractor may be employed by the Contractor.

Reporting

The Contractor is normally required to report monthly on the progress and quality of construction and the supporting activities of ordering materials and designing temporary works. 'Returns' (data) of the number and classes of labour and plant employed on site by the Contractor and sub-contractors are usually also required, together with reports on any accidents and near misses.

Progressing

Assessments of progress should be based upon measuring or estimating the amounts of work completed, rates of work and trends, so as to be able to predict when succeeding activities may start.[3]

Measurement of work done, rate of work and trends may be rough or detailed. Statements of elapsed time alone or man-hours or money expended are of little use, because they give no indication of what has been produced and when things will be completed. Data on progress should then be compared with predictions and the programmes to provide the basis for any action needed to achieve completion dates.

A convenient way of expressing progress is by using indices. Commonly a schedule and commercial index is produced. These are generally functions of earned value. The programme will typically show a planned duration of the works. By comparing the actual performance of construction, values can be attributed to the actual and earned durations. Hence useful indices can be calculated for use when analysing and predicting performance.

Cash flow

Every contractor and sub-contractor who is due to be paid according to progress of work needs to plan his expected cash flow when tendering, and then monitor it during the work. Earlier expenditure or later payment than expected can incur greater costs for a contractor, which can impact severely on his expected profit.

Cost monitoring system

On all but very small jobs every contractor and sub-contractor should have a site cost recording and monitoring system that provides

- accurate reports at regular intervals of the unit costs of all the principal work and overhead charges
- estimates of the likely final costs
- data in the form required for claiming progress payments
- cost data of the completed works to guide future estimating.

This data is needed whether the contract is *fixed priced* or costs are reimbursable. It is confidential to each contractor and sub-contractor unless required for payments under a *reimbursable contract* or part of a contract.

Labour and plant costs depend mainly on good planning, control and discipline on site. All these costs should therefore be recorded, reported and analysed, usually weekly, to show whether these resources were used productively, so that remedial action on problems can be taken before large wasteful costs have been incurred.

Monitoring of the costs of materials is needed to see whether materials are being used economically and to identify and remedy cases of waste and other losses. The accuracy of a costing system is very dependent on information from foremen, gangers, machine drivers and other operators on how man-hours have been used. The definitions of cost items should therefore be clear to them. The way that items of work are listed in a bill of quantities is not necessarily suitable for this purpose.

Each contractor and sub-contractor also needs an internal accounts system which provides reports and forecasts of expenditure, commitments, liabilities, cash flow and the expected final financial state of their contracts. Some or most of the cost monitoring and accounting required may be provided by a regional office or the company head office and will nowadays invariably, be computer based.

Payment to Contractors and sub-contractors

Monthly statements for interim certificates

In most civil engineering contracts the Contractor is entitled to monthly payments for work completed. A statement (valuation) showing the amount due during the month in question is prepared and submitted by the Contractor in a form usually specified by the contract.

The statement should be checked and agreed by the Promoter's Representative who will, after verification, send the Promoter a certificate showing the amount to be paid to the Contractor.

In a traditional admeasurement contract, payment for most of the work done is based on the Contractor's tender rates for each item listed in the bill of quantities. Alternatively, the terms of the contract may be that payment is due when the Contractor completes defined stages of work (*milestones*). This basis of payment is becoming more common, as it is directly related to progress towards the Promoter's objectives. In cost-reimbursable contracts payment may depend upon agreement of the value of progress achieved, not simply the costs incurred. In most contracts the Promoter is obliged to pay the Contractor within a specified period (usually 28 days).

Some contracts allow for neutral cash flow, which means that the Contractor will be given some form of advanced payment based on a forecast of expenditure.

Final payment

The final balance of payment is usually due to the Contractor at the end of the defects correction period. By then the Promoter's Representative should expect a final account from the Contractor and be preparing to check it and issue the final certificate. If there remain any outstanding claims from the Contractor, these should be settled under the procedure stated in the contract.

Claims and disputes

The procedure to be adopted in the settlement of disputes that may arise during a contract is usually established in the conditions of contract. In the ICE conditions, a matter does not contractually become a dispute until the Engineer or an Adjudicator has formally given his decision on it and one or other party to the contract (the Promoter or the Contractor) has formally objected to that decision. It may then be settled by negotiation between the parties (preferably conducted through the Engineer) or, if this fails, by reference to conciliation. If this is not successful, either party can refer it to arbitration.

The most common kind of dispute, where the contractor has objected to a decision of the Promoter's Representative, is a claim from the Contractor for extra expenditure he has incurred, or for a loss he states that he has suffered because of circumstances which the Contractor considers could not have been expected to allow for in the tender. If so, the Representative must decide whether or not there is provision in the contract under which they can properly certify this expenditure for payment by the Promoter.

It is the duty of the Representative and the Contractor's management to record and if possible agree full details of the facts and circumstances relevant to any matter that may be the subject of a claim. Where agreement on facts cannot be reached, separate records should be kept and the reasons for disagreement noted.

The assessment of a claim is a matter for the Representative after discussion between him and the Contractor. Having ascertained all the relevant facts, the Representative has to decide to what extent they consider a claim is justified under the terms of the contract. Having decided this, the Representative should use factual data compiled from the records to price the claim and report this decision to the Promoter and the Contractor.

Dispute resolution

The use of conciliation to resolve a dispute is a term of the ICE and Minor Works conditions of contract. Adjudication is the procedure in the NEC. The results of asking third parties or a law court to settle a dispute are uncertain and the process can be very expensive. They should be used only after failure of the parties to agree by negotiation.

Following practice in the US, where litigation can add more than 10% to project costs, Alternative Dispute Resolution (ADR) procedures are now required in the UK under the Housing Grants, Construction and Regeneration Act 1998, instead of immediate recourse to arbitration. ADR techniques are based upon the use of neutral experts. They require a change in contractual 'culture'. ADR methods vary, and can work successfully only if the parties in dispute are prepared to collaborate in solving a problem.

Communications and records

Communications between the Promoter's Representative under the contract and the Contractor should be in writing, or confirmed in writing.* Every document exchanged between the Contractor and the Promoter or their Representative on the execution of the project – all letters, memos, drawings, sketches, photographs, data on disks, minutes of meetings, progress and other reports, monthly *measurements*, claims and certificates – become part of the administration of the contract, unless agreed otherwise. All these and physical records of site conditions should be kept and filed to form the contract records.[4]

Documents whose importance or usefulness is not wholly clear at the time may become so later, for instance to help with such things as analysing the causes of accidents, failures or deterioration of completed work, the state of buried work, disagreements between the Promoter's Representative and the Contractor over payments or delays, and designing and pricing additional work. The records that should be kept include

- diaries – most importantly, every engineer on a contract should keep a detailed diary of his own work, however insignificant the detail may seem at the time
- notes of oral instructions or agreements
- superseded drawings, which must be kept because something done when the drawing was current may seem very odd several years (and five revisions) later
- an up-to-date health and safety file, to guide all who will work on any future design, construction, maintenance or demolition of the project, and records of safety training and meetings
- as-built drawings – in practice the need to break into old work often reveals that the record drawings are out-of-date or incomplete.

---

* '*Confirmation of verbal instruction*' forms for this purpose are available at the ICE bookshop.

Everybody may agree that good records should be handed over to the users of completed projects. Professional engineers have a duty to make sure that this happens.

References

1. EDWARDS L. S., LORD G. and MADGE P. *Civil engineering insurance and bonding*, Thomas Telford, London, 2nd edition, 1996
2. NEALE R. H. and NEALE D. E. *Construction planning*, Thomas Telford, London 1989
3. WEARNE S. H. (ed.) *Control of engineering projects*, Thomas Telford, London, 2nd edition, 1989
4. CLARKE R. H. *Site supervision*, Thomas Telford, London, 2nd edition, 1988

8

Construction Management Organisation

This chapter describes typical organisations required to plan and control this stage of a project, for both construct only and Early Contractor Involvement (ECI) schemes, where there is one main contractor in a traditional civil engineering contract in the UK.

The Promoter's representative

Traditionally in civil engineering in the UK the Consulting Engineer is the representative of the Promoter from the inception of the project, as described in Chapter 2.

Consultant's Project Manager

The term Project Manager or Design Project Manager is commonly used within consultancy organisations for the *project engineer* with responsibility for the following.

Construct only

- to plan and supervise detailed design, if not already complete, and coordinate the issue of any further drawings to the contractor under whatever procedure is stated in the contract
- to promote and implement Value Engineering and innovative ideas
- to direct redesign if there are varied requirements of the Promoter or changed site conditions
- to estimate the effect which any variations will have on the programme and cost of the project, and to advise on issuing a

Promoter's Change/Variation Order or whatever procedure is stated in the contract
- to advise the Promoter on the progress, trends and likely outcome of contracts
- to manage the project risks
- to manage third party customer care
- to actively manage environmental risks
- to promote a partnering ethos to engender mutual trust and cooperation
- to administer the issue of the certificates for payment to the contractor
- to advise Promoter on Compensation Events, claims and disputes
- to liaise with the Supervisor/Promoter's Representative on all the above.

Under the NEC Form of Contract the 'Project Manager' is named within the contract and is appointed by the Promoter to manage the contract on his behalf. The contractual role of the Project Manager is defined in terms of the decisions and actions he has to take. The equivalent role in ICE Conditions is 'The Engineer'.

Delegation of authority

Under the NEC conditions of contract the responsibility for most communications with the Contractor is usually delegated by the Project Manager to the Supervisor. The equivalent role in the ICE Conditions is the 'Engineer's Representative'.

The extent to which the Project Manager can delegate their powers in a contract is usually limited in the contract. The Project Manager should inform the contractor in writing of the extent of delegation of their powers.

The Supervisor

The role of the Supervisor under the NEC conditions of contract is to check that the project is constructed in accordance with the contract. The principal duties are

- to carry out the duties delegated by the Project Manager
- to check that the contractor has organised their work to achieve the accepted or approved programme
- to examine the methods proposed by the contractor for the execution of the project, the primary object being to see that they should ensure safe and satisfactory construction
- to ensure that the Contractor complies with the requirements of the Project Health and Safety Plan
- to assist the Contractor to interpret drawings and understand the specification, and refer questions to the design project manager/project engineer
- to supervise the project to ensure that they are being executed in line with the requirements of the Project Quality Plan
- to assess and record the progress of the work in comparison with the programme
- to execute and/or supervise tests carried out on the site, and inspect materials and manufacture at source
- to keep a diary constituting a detailed history of the work done and all events at the site and submit periodic progress reports to the Project Manager
- to advise the Project Manager on the monthly assessments of the amount due to the Contractor
- to agree and record the relevant facts for any work or event for which the Contractor may claim additional time or payment
- to direct the production of as-built drawings and the health and safety file
- to manage risk by advising the Project Manager on potential problems in good time for them to be avoided or their effects minimised.

The Engineer's site team

Except on small projects, the Supervisor usually leads a team of assistant engineers, inspectors and support staff. Their actual numbers and organisation depend upon the size of the project, the variety of the work and distance from head office or services.

The role of the Inspectors is to supervise the Contractor's work, for instance the mixing and placing of concrete and any such work requiring constant supervision. The duties of Inspectors demand practical experience, objectivity and tact in order to gain the respect of the foremen and skilled workmen employed by the Contractor. Corresponding roles are needed in other contract arrangements, for instance where a project is designed and supervised by the Promoter's own staff.

Early Contractor involvement (ECI) or construct only

The Promoter's Representative role is determined by the type of contract and the stage at which a contractor is engaged to take responsibility for the project development (early engagement is referred to as Early Contractor Involvement (ECI)). Until the Contractor is procured, the Promoter's Representative responsibilities cover the following

- develop the design within the required scope and brief ensuring compliance with CDM Regulations
- manage the scheme development through the planning or statutory processes
- achieve a satisfactory outcome at any Public Inquiry if required
- assist the Promoter to define the budgets and programme requirements
- manage consultations with stakeholders, the public and other interested parties
- ensure that effective communication between contractor and the Promoter.

The Promoter may wish to procure the Contractor early at which point some of the responsibilities above may be transferred to the Contractor. These will be determined within the contract.

During construction, the role of the Promoter's Representative is greatly reduced from the role previously outlined for 'construct only' schemes. Their main responsibility is to monitor the following

- the progress of the contractor and keep the Promoter fully informed
- deal with scheme changes and events on behalf of the Promoter
- determine the payments to be made to the contractor based on work completed.

Main Contractor's organisation

Figure 13 shows an example of a contractor's structure of company departments and the main responsibilities for managing contracts. All roles under the Operational and Commercial Directors are responsible for project delivery. However, these teams are supported by functions that are located off-site in the company's head office or a regional office.

Figure 14 shows what might be a contractor's management structure on site for a medium-size project. Here some roles are specialist ones in a function or discipline, for instance planning. Others are in a section or area of the site. The nature of these duties varies significantly, depending on the variety of work, size and layout of a site and the terms of the contract. On major projects there tend to be more specialists; on smaller and traditional contracts there tends to be a wider range of responsibilities for each individual.

Contractors' project managers and agents

Contractors' project managers are usually experienced engineers. Most of the financial risks of a construction contract are on site and so the contractor's project manager is usually given wide powers by their company to plan and control the work. The project manager's main role is to successfully deliver the scheme to time, cost and budget, which includes responsibilities for the following

- construction
- health, safety and environment

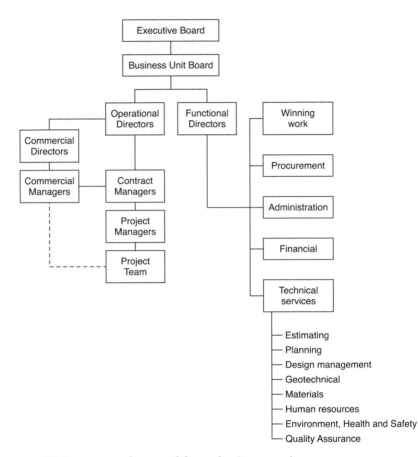

Fig. 13 Structure and responsibilities of a Contractor's company

- compliance with the contract
- the commercial success of the contract
- management of the Contractor's site staff
- programme management
- liaison with the Supervisor
- stakeholder liaison and communication

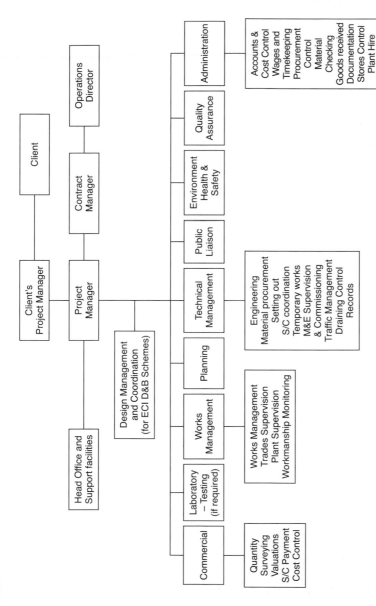

Fig. 14 On-site Contractor's management structure

107

- risk management
- performance management
- advising the Promoter on scheme budgets.

Sub-agents

On larger sites areas of the work are usually the responsibility of sub-agents. Depending on the size of the particular project they will have varying numbers of staff, principally section engineers, assistant engineers and inspectors. Sub-agents' main responsibilities are

- day-to-day site management
- health and safety
- industrial relations
- management of sub-contractors
- productivity and workmanship of the plant and labour.

The use of labour, plant and transport is usually supervised by the general foreman and sectional foremen, depending on the size, variety and spread of work.

General foreman

The general foreman is the link between the management and the foremen and gangers in direct charge of labour. Their personal influence on the site organisation can be a strong factor in achieving and maintaining efficiency.

The general foreman's main responsibilities are

- allocation of labour to site operations
- supervising flows of materials and the management of stores/compound areas
- motivation of the labour force
- site communications between all foremen and gangers
- site safety – including workforce briefings, statutory inspections and site tidiness.

Chief Engineer/Engineering Manager (on large projects)

The Chief Engineer is responsible for the technical methods needed to ensure the quality and accuracy of the works, through guidance of the section engineers and all the Contractor's staff. They are responsible for any design needed on site, especially of temporary works, and will draw on head office engineering and research departments where appropriate.

Section engineers

Section engineers usually have an appreciation of design, construction, health, safety and environmental requirements. They are ultimately responsible to the Project Manager/Agent, but are usually directed by the sub-agent. On large or technically complex schemes they will obtain assistance and technical guidance from the Chief Engineer. They report progress and measurement/scheme change issues to the quantity surveyor. Each section engineer must liaise with the foreman in their section, to plan the work to be executed daily, weekly and monthly.

Cost engineer or quantity surveyor

On larger projects a separate cost engineer or quantity surveyor and assistants may be needed to record the quantities of work done, manage sub-contractors' payments and prepare the information for the Contractor to request payment as defined under the contract. Interim and final measurements have to be substantiated and agreed with the Supervisor.

The quantity surveyor will notify the Supervisor of any events or changes in work scope and they will be responsible for evaluating and agreeing the effect of change. If agreement of these variations is not achieved, the quantity surveyor will assist in the dispute resolution procedure defined in the contract.

Normally the quantity surveyor is also responsible for managing correspondence and instructions on the main contract and sub-contracts, and advising the Contractor's project manager/agent on contract and sub-contract matters. On smaller projects these tasks are part of the job of sub-agents or section engineers, supported by a visiting quantity surveyor.

Planner

The planner is responsible for the development of the overall contract programme in conjunction with other key members of the project team. They will update the overall contract programme with progress to date at intervals specified in the contract or as required by the Project Manager. They will also usually provide the site team with a weekly or fortnightly programme giving the work elements for the forthcoming month.

It will be either a full- or part-time role, depending on the size and nature of the scheme.

Quality/safety/environmental coordinator

The role of the coordinator is to develop, manage, implement and regularly monitor the requirements of the Project Quality Plan, Health and Safety Plan and Environmental Management Plan. These roles are carried out by one person or by separate individuals, depending on the size of the project. Many contractors now have an integrated management system which combines the business processes for all these three areas.

Public liaison officer

For nearly all projects that require the specialist services of a public liaison officer, the duties can be described as follows.

- Consultation and liaison with affected landowners, residents and local businesses. Assisting at key liaison group meetings. Communicating back to the project team.
- Provision of a regular surgery or hold public exhibitions for the public for general discussion, raise issues, etc.

- Respond to compliments and deal with complaints.
- Control and organisation of the media, local radio, leaflet/letter drops on progress, town notice board, dedicated e-mail address, traffic switches, etc.
- Liaison with and organisation of trips and visits by local schools and interested parties.

The primary function is to ensure that the public perception of a project remains positive.

Support functions

Most contractors will have head office support functions as indicated in Fig. 13. They are responsible for both support to the projects and for ensuring a consistency of standards and approach across the organisation.

Appointment of site staff

If the Works are to start well and proceed economically, much will depend on the early appointment of the Contractor's project manager/agent and their staff so that they can begin their planning and other preparations and establish selection, induction and training facilities before labour arrives in any numbers. There is a risk that these preparations will be inadequately planned if one person has to do several others' work during the initial period.

Selection of adequate and experienced staff and briefing them on the project, its priorities, risks and organisation are particularly important if the site is remote – and especially if overseas.[1] Ideally key members of the team should be involved with the project at *bid* stage to facilitate continuity, however in all cases a formal handover to the site staff should be carried out.

Industrial relations – communication and procedures

The *Working Rule Agreement* for the Construction Industry (Construction Industry Joint Council), negotiated between the Promoters' representatives and trades unions, contains the terms of employment and grievance procedures for manual workers in the civil engineering industry.[2] Increasingly contractors and sub-contractors in the UK employ skilled and other labour as individual sub-contractors rather than as permanent employees, as this is more flexible for the Promoter and can have tax advantages to the employees, but the Working Rule Agreement provides a basis for their employment. Thorough knowledge of this agreement and the amendments made to it from time to time is therefore needed for successful industrial relations.

Civil engineering in the UK has comparatively good industrial relations and care should be taken to maintain them. Good industrial relations on site are achieved by the management being seen to be consistent, fair and reasonable. To this end a good communication system should operate. Management should always be prepared to meet employees' representatives, resolve factual questions and explain policies.

Suitable provisions should be made for foreign/migrant workers, as required.

Incentives

Contractors use incentive bonus schemes to try to achieve good productivity from manual employees. The basis of a good incentive scheme is that it should give a person of average ability the opportunity to earn more than their basic wage in return for increased production. Incentive schemes need to be seen to be fair. They require both technical and psychological skill to formulate and apply, otherwise discontent can quickly arise. Weekly measurement of production and the calculation of bonuses requires promptness and accuracy.

Production targets and the bonus applicable to them should be clear and agreed between a contractor or sub-contractor and the representative of their employees, and thereafter altered only if circumstances justify changes.

Not all work can be made the subject of a bonus by the direct measurement of output.

Employees who provide support services to those on bonus targets should therefore be given a financial interest in the work which attracts bonuses and so gain some benefit when bonuses are earned by their colleagues.

Site office administration

An administration/office manager (titles vary) on larger or remote sites is usually responsible for secretarial and other administrative services to the site organisation, the payment of wages, sickness records, minor purchasing, and the checking of the receipt and safeguarding of materials. On smaller and urban sites most or all of these services are provided by the Contractor's head or regional offices.

Sub-contractors' organisations

Sub-contractors' organisations are generally similar to those of main contractors, but smaller and more specialised to suit their scale and range of work.

Many projects are now carried out in a partnering environment in which major sub-contractors form part of the integrated project team.

Early Contractor Involvement (ECI) or design and construct

In some circumstances the Promoter employs the contractor to both design and build a particular project.

The extent of the contractor's design responsibilities will depend upon the point in the process at which the contractor is employed and it will be defined within the contract.

Some of the Promoter's representatives' responsibilities outlined previously for construct only will consequently be transferred to the Contractor. The Contractor employs a designer to complete the design and supervision on the Contractor's behalf.

In this instance the contractor's designer is responsible for

- developing a buildable, cost effective design solution within the required scope and brief
- obtaining all the necessary approvals and certification
- ensuring the design is completed within the cost and programme constraints
- developing the Environmental Management Plan taking into account all environmental constraints
- dealing effectively and in a timely manner with scheme changes
- site supervision and auditing the contractor's site quality records
- assisting the contractor in updating the scheme budgets.

Design manager

For these schemes the Contractor will employ a design manager/design coordinator to manage the interface between the Contractor and their designer. Key duties include

- providing a clear design brief
- assisting the designer with buildability, value engineering and designing out risks
- agreeing the design programme and planned resources
- regularly reviewing progress, costs to date and forecast costs with the designer
- effectively managing design change
- liaising with the site team on preferred solutions and design development.

References

1. LORAINE R. K. *Construction management in developing countries*, Thomas Telford, 1991
2. MARTIN A. S. and GROVER F. (eds), *Managing people*, Thomas Telford, London, 1988

9

Testing, commissioning and handover

Planning and organisation

The period of testing, commissioning and handover of a project is the transition from its construction to its occupation by the end users. For the simplest projects this stage may consist of only a formal handover. For projects which include mechanical or other operating systems, controls and equipment, this stage is a separate set of testing and commissioning activities which overlap their supply and installation.

Planning

Testing, commissioning, handover and occupation requirements should be incorporated in planning activities from the earliest stages of a project, so that provision for them can be made in design and in deciding contractors' responsibilities. Together with the necessary budget provisions, this planning should be part of the project strategy. Suitable provisions can then be included in the relevant contracts with regard to responsibilities for testing, commissioning, partial and full handover, including commissioning by both the Promoter and the Contractor. This may need to allow for continued construction or fit-out work within a partially operational system or building.

Detailed planning of testing and commissioning activities will usually be necessary in the pre-commissioning stage, in parallel with construction.

Organisation and resources

The Promoter may need to establish a commissioning and occupation team which includes their own representatives, others who are to be the eventual owners, users or occupiers, the engineering team and specialist commissioning personnel.

The resources needed should be identified and procured early in the project. An effective and active management structure should be established under the Project Manager, who should ensure that the responsibilities of each party are clearly defined and planned in order to ensure the smooth commissioning and handover of the project.

Health and safety

The testing, commissioning and operation of equipment and systems in a completed facility may include hazards with which construction managers and workers are not familiar. Other personnel from various organisations are likely to be working closely together, many not familiar with site hazards.

Health and safety responsibilities and the transfer of the control of hazards, through the handover of the health and safety file, from construction to commissioning and operations staff should therefore be clearly defined and effective permit-to-work and other control procedures established. Detailed studies of risk and responsibilities, and training and induction in health and safety for all personnel during this stage are required.

Public liability

The safety of the public, as well as of occupiers and users should also be considered at all stages of commissioning.

Inspection and testing of work

Inspection and testing

During construction and on completion of parts of a project, inspection and testing are usually required in order to confirm compliance with the drawings and specification. Testing and inspection are generally required for static components of a project, while dynamic components such as machinery require testing and commissioning. The Project Manager or the Engineer under the ICE 6th edition, supported by their team, is usually responsible for inspections on and off-site and for testing of materials.

Test criteria and schedules

The performance tests and criteria to be applied to any aspect of work should be specified in the contract for that work as far as possible, so as to enable the identification of the state at which an acceptable quality or degree of completion has been achieved. Depending on the type of project, samples and mock-ups or factory inspections and acceptance tests may be required. If they are, the responsibility for their cost should be defined.

Schedules (lists) of the necessary inspections and tests should be agreed through collaboration between the Project Manager and all parties.

Commissioning

Commissioning roles

Commissioning should be the orderly process of testing, adjustment and bringing the operational units of the project into use. It is generally required where systems and equipment are to be brought into service following installation.

Commissioning may be carried out by the Promoter's or Contractor's staff, by specialist personnel or by a mixed team. For complex industrial projects, a commissioning manager is usually appointed by the Promoter to plan the commissioning, preferably early in the project, establish budgets, lead a commissioning team and procure the other necessary resources. For simple projects, commissioning is usually undertaken by the Contractor, subject to the approval of procedures by the Promoter or their representative under the contract. The commissioning of familiar or small process and industrial facilities is usually carried out by the operator with the advice and assistance of the suppliers of the equipment and systems.

Organisation and management

Whatever the contractual relationships, the commissioning and occupation of a large project can involve a number of parties and risks. Active management, clear procedures and effective communication are thus essential. The management organisation for pre-commissioning and commissioning should be agreed, and key personnel mobilised, during the construction stage, and organised so as to provide a

- simple structure to suit the specialised nature of the work, with clear responsibilities and without multiple layers of management
- single point responsibility for all commissioning activities at all times, under the direction of a commissioning manager
- rigorous procedures in respect of health and safety.

Commissioning process

The commissioning process may need to allow for continuing construction alongside commissioning activities and, in many cases, occupation of premises or operation of completed systems. Thus, while construction proceeds on the basis of physical site areas and technical specialisations, commissioning usually has to be undertaken on complete operating sub-systems and units.

Commissioning schedule

For all but simple static work the commissioning manager should draft a schedule listing all the items to be commissioned, their interdependence and the standards of performance to be achieved.

Commissioning is usually carried out progressively at the levels of sub-units, units, systems and the whole project. The schedule should provide for these activities being carried out sequentially, alongside continuing construction. Contingency plans should also be prepared, to allow alternative commissioning sequences if problems are encountered.

Commissioning plan

The commissioning plan should include a programme and identify all activities and the necessary resources and procedures. These should include the supply of power, water or other services for testing, materials, consumables, spares, labour and specialist expertise. The plan should also identify procedures for managing emergencies and for rectifying defects, both before and after handover.

Staffing and training

In order to ensure that the project is put into operation rapidly, safely and effectively, all the commissioning, operating and maintenance personnel should be appointed, trained and briefed before commissioning starts. This needs to be planned from project inception, so that the roles and activities of the commissioning and operating staff are integrated into a coherent team to maximise their effectiveness.

Completion and handover

Practical and sectional completion

In most construction contracts there is provision for partial or stage completion and handover of sections of the project. Once the contractor considers they have completed the scope of work within their responsibility and fulfilled all the necessary obligations relating to a section or the whole of the project, they may apply to the Project Manager or the Representative under the contract for a completion certificate.

For many projects, sectional completion may signal the handover of a physical unit from one contractor to another for further work, fit-out or equipment installation. Well-defined procedures for such handovers should be established and agreed well in advance and should preferably form part of the relevant contractors' contracts.

Defects

Before acceptance of the works and issue of a completion certificate, the Project Manager or the Representative under the contract should inspect the relevant works jointly with the Contractor and prepare a list of outstanding items of work or defects. Together with a programme for completion of the work, the schedule should be agreed with both the Contractor and the Promoter or a follow-on contractor, as appropriate.

Documentation

The commissioning and handover state is the point for finalising the project documentation. There are three categories of documents to be handed over by designers, equipment suppliers and contractors

- records of the equipment and services as installed
- commissioning instructions, including safety rules
- operating and maintenance instructions, including safety rules.

Generally the first category will include design and performance specifications, test certificates, defects lists, as-built drawings and warranties. In addition to these, documents for commissioning and operation will include permits, certificates of insurances, operation and maintenance manuals and handover certificates.

Acceptance, handover and certificates

Following achievement of the relevant performance test criteria, the Project Manager or the Representative under the contract is normally required to issue a certificate of completion, or partial completion as appropriate, to the Promoter. This is accompanied by the relevant outstanding work schedule and completion programme. For complex projects, a handover certificate and detailed handover procedures may be used. Under many forms of contract, the issue of a completion certificate allows the release of part of the retention money held back by the Promoter from payments to the contractor.

Warranties and defects liability

In the UK most of the contractor's contractual responsibilities and general liabilities in respect of the works handed over pass to the Promoter (or to another contractor following on to do other work) upon issue of the completion or handover certificate. Thus, warranties for equipment, insurance liabilities and responsibility for operation, day-to-day cleaning, maintenance, health and safety may pass to the Promoter or follow-on contractor, as appropriate.

In many other countries, contractors and suppliers of equipment continue to have legal requirements to be insured against public liabilities.

Occupation

Planning

The occupation, on the issue of a completion certificate, of a major development may, in itself, represent a significant project, requiring extensive planning and development, particularly for large purpose-built commercial facilities.[1] The occupation sequence may differ considerably from construction, in that it centres around employees themselves and involves both the style of management and the culture of the user's organisation.

Consultation with representatives of users and employees can thus be important in project planning and design to identify their requirements and what facilities they require. Their involvement can also be important to ensure their commitment to the results and hence the ultimate success of the project.

121

Organisation and control

To ensure a successful occupation sequence, it is not uncommon for the user to appoint a dedicated occupation or 'migration' project team, headed by a project manager. Their responsibilities will include planning, programmes, budgets, methodologies, contingency plans, health, safety and control procedures and the identification of risks. Steering groups and representatives' groups may also be established to promote consultation and communication and assist with identification of requirements. The team will need to arrange support services and utilities and may also need to coordinate migration with continuing construction or fit-out work. They may also need to reconcile different interests of owners and occupiers.

References

1. CHARTERED INSTITUTE OF BUILDING. *Code of Practice for project management*, Ascot, 1993

10

Operation and maintenance

Operation and maintenance needs

A completed project becomes part of an operating system or public facility after handover. Some structures or facilities may then need only regular inspection and maintenance, but more often become part of a wider group of the Promoter's assets which, in order to deliver the Promoter's original requirements, need regular and frequent operational and maintenance intervention. Individual infrastructure facilities, for example, a road, bridge or hospital, require not only inspection and maintenance, but also planning of their use in a wider highway or healthcare system. Temporary or partial closure for maintenance, refurbishment or the replacement of components can affect traffic flows or healthcare provision across a wide area.

The delivery of an asset available for a promoter to use on a continuing basis requires careful consideration of both operational requirements and appropriate maintenance regimes at an early stage in the gestation of the project.

The annual cost of operating and maintaining infrastructure or facilities varies significantly and, depending on both design life and discount rate, may represent a similar order of magnitude as the initial capital cost on a net present value basis. Operations and maintenance regimes therefore need to be considered throughout the development of a project.

The costs to operate and maintain an asset or facility in use clearly requires the economic consideration and understanding of the whole-life cost of that asset. Specifically, the design, construction, operation, maintenance, decommissioning and, in some cases, financing of the project need to be fully understood throughout various stages of a project's development. To ensure the asset remains in the appropriate condition upon completion and throughout its operational life, critical or sensitive elements of the asset will require periodic inspection to enable the assurance of the continued structural integrity, serviceability and fitness for purpose of the asset. The inspection regime will itself be in part dependant upon both the planned and actual operation of the asset and the extent to which the envisaged maintenance regime is realised.

The optimal whole-life cost will further depend on financial assumptions including, *inter alia*, demand forecasts over time and discount rate chosen and may be further constrained by initial affordability considerations.

Planning for operation and maintenance

Economic considerations

The costs of operation and maintenance should be considered from the inception of the project, as mentioned in Chapter 1 and discussed above.

Differences in the costs between alternative schemes should influence the choice between them. Failure to consider these costs may lead to a choice of design that is uneconomic (or unsafe) to maintain, operate or decommission. The health and safety of operations, consumers and third parties throughout the life-cycle of the asset, including its operation, maintenance, refurbishment/reconfiguration and decommissioning is clearly both a necessary and regulatory requirement.

Operational and maintenance assumptions also affect the initial capital cost. In a water treatment facility, for example, the choice of ozone instead of chlorine as a disinfectant may increase the construction cost but reduce the operating cost.

Non economic considerations

In addition to the consideration of whole-life economic costs and benefits, Promoters are increasingly sensitive to the environmental foot print of their assets. Beyond an initial 'Environmental Impact Assessment' (EIA), promoters are often considering the embodied energy of completed assets, the quantity of CO_2 generated by the asset throughout its operation, the quantity of other consumables used throughout the assets life and any legacy issues that might remain on decommissioning. While consistent approaches to these issues have yet to be formalised at the time of publication they are being seen as increasingly important considerations.

Flexibility

For projects with extended in-use life expectancy, it may be prudent to consider the effects of operational and maintenance change and to assess the extent to which the project may be able to accommodate changes. Refurbishment and reconfiguration are often used to adjust or enhance capacity utilisation. While it is not possible to accurately predict the future, opportunities to allow flexibility in both asset operation and maintenance are often valued by promoters.

As-built drawings

As-built drawings should be prepared during construction, while detail can be seen and checked. For civil engineering work this is often undertaken by the Promoter's representative and their staff. For electrical, mechanical and process projects or elements it is frequently the contractor or a sub-contractor who prepares the drawings and submits them for approval by the Promoter.

At the end of construction, not later, a fully detailed set of drawings should be available to the operators, to represent the works as-built. They should include modifications made during the defects correction period. The drawings will be used to plan the use and control of the facility, maintenance and repair work, and as the basis of further design work for further development of the facility. The drawings handed over to the Promoter at the completion of the contract must therefore be a reliable representation of the actual asset at that time.

Manuals and operating documentation

The drawings and other documents must specify the operation and maintenance of hardware (equipment and structures), software, and the supporting needs of planning, stock control, billing and revenue collection, customer and employee relationships, training and career development.

Contracts for operation and maintenance

Increasingly the Promoters of projects are employing other organisations to manage the operation and maintenance of completed facilities. These 'facilities management' (FM) contracts may be separate from those for design and construction, employing a following-on services contractor, or they may be part of a comprehensive contract for the project.

Process facilities, such as a water treatment plant comprising a reservoir, treatment facility and metered distribution network, require every component to have well-defined contracts for their operation. The operational resources required, in the form of power, chemicals, manpower and the collection of revenue, can result in complex operation and maintenance programmes.

Figure 15 shows a procedure for defining the need for a maintenance agreement with a sub-contractor who has supplied and installed equipment for a project.

Increasingly promoters are integrating the obligations for operating and maintaining the asset with the contractor who supplied most or all of a project. The contractor should then have a greater incentive to provide a facility which will be of appropriate quality and operate effectively for the prescribed number of years, and ensure that skilled personnel will be available to train the Promoter's operating staff before transfer. The scope of these contracts can include many different combinations of operation, maintenance, condition on handover and, in some instances, training.

Operation only

Under these contracts the operator undertakes to operate and manage the facility, with any maintenance being the responsibility of the Promoter. A toll bridge, leisure complex or local road network could be examples where the operator would be primarily concerned with ensuring the asset is available to function on a day to day basis but who may not be responsible for the associated maintenance regime. The cleaning of an office, the provision of helpdesk, security or porterage services are examples of individual 'soft' FM services. These may be procured by the Promoter on an individual service basis or bundled together to provide a single operational support interface.

Maintenance only

In this case the operator, usually the contractor, is responsible for solely the maintenance of the facility. Examples may be the maintenance of particular equipment which may be required due to wear and tear over a fixed period, for example, the regular service of air handling fans or replacement of associated filters. Heating, ventilation, air conditioning, electrical and building fabric maintenance services are generally referred to as 'hard' facilities management. Similar to soft FM, hard FM services may be procured individually but are more often bundled together.

Operation and maintenance only

Under these *terms of contract*, the operator undertakes to operate the facility and perform all routine and non-routine maintenance necessary to sustain the facility in full working order and to ensure the

127

Civil engineering procedure

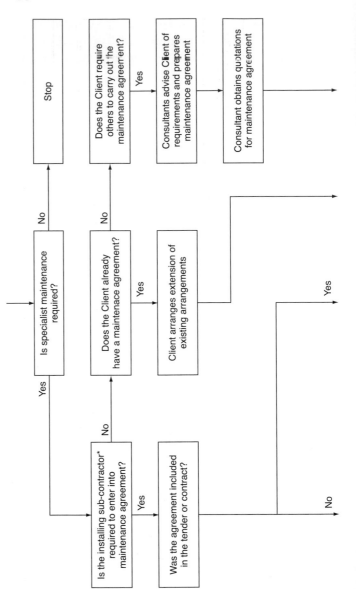

128

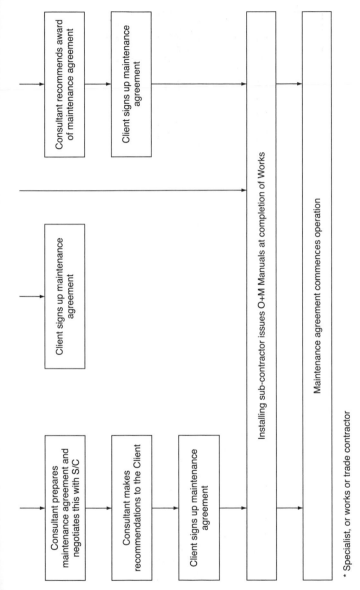

* Specialist, or works or trade contractor

Fig. 15 Procedure for defining the need for a maintenance agreement

129

facility is handed over at the end of the contract in accordance with a pre-agreed condition state. In addition the operator is responsible for the provision of spares and consumables. An example may be the operation and maintenance of a water treatment facility or the provision of all but clinical services in a hospital.

Operation, maintenance and training (OMT)

Here the operator is obliged to train personnel, usually the Promoter's, to operate and maintain the facility until transfer. This form of contract may be considered for process or industrial plants which require high levels of operation and maintenance skills in order to ensure that both revenues and costs modelled in the promoters feasibility assessment are capable of being achieved, is generated during and after the contract.

Conditions of contract and risk

These long-term contracts thus differ significantly from construction contracts in their scope, duration, supervision, method of payment and the nature of risk taken by both the promoter and the contractor. The extent to which risk is borne by the party best able to manage that risk will determine the delivery of optimal value for money. In aggregate, the balancing of initial capital cost, annual operating costs, planned preventative and reactive maintenance costs, together with the expected life of the asset, discount rate and decommissioning costs leads to the optimal whole lift cost of the asset and the ideal allocation of risk.

Integrated contracts

The proliferation of build, own, operate and transfer (BOOT), design, build, finance and operate (DBFO) and many associated variants has accelerated in the UK, primarily driven by the procurement of Private Finance Initiative (PFI) or Public Private Partnership (PPP) projects by the government. In the majority of cases the broad intent is for the Promoter to procure the delivery of services for a period of time from a contractor rather than to procure or own a particular asset. The Contractor in this case is often referred to as a Special Purpose Vehicle (SPV) and typically includes a consortium of organisations who together have the ability to design, construct, operate, maintain and fund the project for a given concession period, e.g. 25 years. At the end of the concession period, the asset is handed back to the client in a pre-agreed condition state.

Training of operations personnel and managers

Training of operatives and managers may be part of an operation and maintenance contract. Training has two elements, the act of communicating knowledge and the act of receipt of knowledge. In many instances the latter is unfortunately disregarded as it can be difficult to quantify and assess in legal terms.

The first step in training must be to assess the existing skills of those who are to be trained. The training contract should clearly define how this should be achieved and against what criteria skills will be measured – in terms of quality as well as quantity. Trainees' effectiveness should be evaluated. The contract documentation must provide an appropriate mechanism for this.

The educational objectives of all training programmes should be specified in the contract. They should specify what the trainees will be expected to be competent to do, rather than what they should know. Job descriptions must be specific and unambiguous, terms such as 'awareness' and 'understanding' should be avoided.

Training should not be considered to be an appendage to a contract designed to perform a dissimilar function. If training is to be included in the operation and maintenance contract, the human and material resources that will be allocated to training need to be defined. For the trainer, skills in interpersonal communication, sensitivity to different cultural norms and values, and knowledge of training methods are as important as technical competence.

131

Overseas operating contracts

Standards of operation and maintenance of many projects in both developed and developing countries vary considerably. Common problems include the use of inappropriate design and technology, inadequate organisations, and the low priority and status promoters may give to maintenance work.

Care must be taken to assess the factors which may affect maintenance on overseas projects. Difficulties may sometimes arise due to the cultural differences and/or scarcity of trained managers, technicians and craftsmen, a shortage of foreign exchange or interruption of supplies, power, spares or raw materials. The division of responsibility for these factors needs to be clearly defined in a contract.

11
The future

Change

The fact that this is the sixth revision of this guide since its first publication in 1963 indicates how rapidly changes occur in the construction industry. Indeed the pace of change is accelerating in many countries. The previous revision of this book mentioned a variety of alternative methods of procurement and contract procedures being introduced into civil engineering. All of these are now in use and have therefore been described in this revised edition.

The industry is likely to continue to change and its future procedures will be determined by further developments in

- UK, EU and overseas markets
- public and private financing of projects
- technology
- European and world political changes
- environmental requirements
- conditions and methods of employment.

The professional engineer has to learn to deal with changes and to use them positively. Changes are opportunities for the advancement of organisations, the profession and the individual.

Changes in markets

Promoters' demands

Governments in the UK and other countries have introduced a

variety of procurement methods to try to attract private investment for the construction of new and improved transport systems and other infrastructure. More and more countries continue to adopt this method of procurement.

The privatisation of former public services and nationalised industries means that they cease to be state agencies when promoting civil engineering projects. As commercial companies they have different demands, aspirations and pressures from their shareholders and their customers. This changes their trading patterns and requirements and as a result many have introduced alternative and innovative methods of procurement.

Public demands

Future needs for new projects cannot be predicted simply by extrapolating past growth and as this new edition is published; there is an unprecedented construction boom in many parts of the world. For instance, road construction and improvements already form a large market for civil engineering. Figures from the Highways Agency indicate that road travel could increase by a third by the year 2025, but the public are unlikely to agree that this trend should be answered only by constructing more roads. Society and industry will demand more coordinated road, tram and rail developments. Lack of public and private finance may limit new construction and instead electronic tolling and other disincentives will be sought to apply to motorways and other systems, particularly at busy times. Integrated information systems for journey planning and traffic management are becoming normal. Future transport policies will thus change the nature of new infrastructure projects. Civil engineers can expect substantial roles in planning, designing and constructing them, but with greater need for inter-disciplinary skills and cooperation in making complex decisions.

The ageing population is creating an increasing demand for new leisure and retirement facilities. The design and construction of these has created a major market.

International markets

Maintenance of the existing infrastructures in the developed world is a large volume of work. Future projects in these countries may be mainly extensions and improvements to existing facilities and smaller works for local social and economic reasons. Maintenance and upgrading of these facilities is increasingly becoming the province of PFI to divert the burden from government exchequers.

The former eastern European bloc countries still need much investment in their energy, transport and services infrastructure. They continue to seek financial and other support from Western Europe. Those countries have considerable construction expertise, and as their economies improve they will require more technical and managerial expertise from the West.

The developing countries will continue to be a source of construction work, with major projects financed by the international banks and aid agencies. Most countries in Africa, South East Asia, South America and much of the Pacific Rim continue to seek more foreign investment.

China is proving to be a massive market, but generally only seeks technology transfer and some services from the West as China looks to export its construction abilities to Third World countries. India and Pakistan have well-established construction industries and also mainly want technology transfer.

The industry

Competitiveness

Many factors can affect the successful execution of projects that satisfy customers. Of prime importance is delivering value for money. This demands innovative and excellent engineering, whether undertaken by the Promoter, consultant or contractor, supported by safe and economic construction methods.

Designers and contractors are under increasing pressure from promoters and their financial supporters to reduce the costs of new projects and other work, both in the UK and other countries. To achieve this everyone in the industry must continue to strive for improved efficiency in design and construction. The means of achieving this include

- best use of time, by individuals at all levels
- strategic use of information technology (IT), for predictions, modelling and managing design and construction as a manufacturing process
- greater use of systematic techniques for project selection, risk management and planning of the use of resources
- the use of newly developed materials
- further standardisation and prefabrication of components for systems and structures both to increase economic production and to avoid weather susceptible activities on site
- mechanisation of more labour-intensive site operations
- improved plant design and materials handling
- ongoing development of technology such as satellite-aided surveying
- unified design, contract and construction data systems
- better use of value engineering and quality assurance systems
- much greater training and development at all levels, in technical, managerial and personal skills
- the encouragement of new generations into the industry to combat skill shortages
- adoption of a continuous improvement culture and more systematic learning from project successes and failures

and of course continuing research, development and innovation.

Health and safety

Inevitably sites are areas of risk. In the UK the law now forces all parties to anticipate and minimise these risks. The legal liability of promoters and designers to initiate health and safety programmes has greatly influenced the health and safety attitudes and procedures of all parties from project initiation to operation.

Similar legislation applies across the European Union and most industrialised countries. Standards and their enforcement vary greatly from country to country, but professional obligations and in many cases new legal requirements to improve standards have to be allowed for in planning future work.

Organisations

The efficiency and competitiveness of the civil engineering industry are increasingly important to individuals, companies and governments. To survive and succeed engineers and their employers have to give more systematic attention to how people and their work are structured and controlled, within organisations and in contracts between them.

Organisations and their systems should be designed to be effective and motivating, not automatic or bureaucratic. Within an organisation, the procedures and techniques which are appropriate vary from project to project, and are different for promoters, consultants and contractors. The professional engineer should be aware of the needs and be prepared to decide how the alternatives available should be applied to suit each project.

Implications for consulting engineers

The roles of consultants as specialists, experts, designers and project managers will continue and expand owing to more sophisticated design and construction methods. Increasing use of the NEC contract system and greater reliance on contractors' quality assurance have changed the site role of the consulting engineer from supervisor to technical consultant.

Promoters increasingly want one organisation to be responsible for all that is needed for a project. The effects of this are

- greater use of design and construct contracts, including design-build-finance-operate contracts for public projects in the UK and other countries

- Early Contractor Involvement (ECI) contracts
- more multi-disciplinary consulting engineer organisations in order to offer the range of skills needed to compete for employment on these projects
- mergers and acquisitions among consulting firms in order to provide the scale of resources needed for the larger projects and bear the increasing risks of competition.

Some promoters and contractors have acquired consulting engineer firms to be their in-house designers and project managers, while others have developed their own design and management teams.

These trends are likely to continue. In addition, the UK and other governments have sold their technical departments or establish them as consulting companies.

Implications for contractors

Pressures similar to those felt by consultants are leading to further reduction and rationalisation amongst contractors. In response to the wider range of procurement arrangements demanded by promoters, many contractors are becoming project managers as well as constructors. They have to develop multi-disciplinary teams capable of promoting, designing, managing, constructing operating and maintaining projects.

Contractors from other European countries are now established in the UK. This trend will continue, further increasing the competition for contracts. To achieve the economies of scale and the financial strength necessary to compete in UK and overseas markets, UK-based contractors may have to work more in joint ventures or combine by merger or acquisition. This may be with other local contractors or with those overseas to get better access to other markets. It will lead to the introduction of different management and procedural systems.

A much more complex industry is developing. Organisations and the responsibilities and authority of staff are becoming significantly more flexible and responsive to different priorities of projects.

Management structures in many organisations are flatter to improve communications and reduce costs. These trends can greatly affect individuals and organisations, and there is no reason why they should reverse.

Contracts

This guide describes the number and variety of general conditions of contract available in the UK and internationally. The problems that this produces are widely recognised. There is now substantial agreement that adversarial types of contract should no longer be used, resulting in a trend towards a single family of contracts for all construction. Notwithstanding, many of the larger complex projects found in today's marketplace are administered under bespoke contractual arrangements, particularly where they bring finance to the contracting package as in PFI and PPP contracts.

There will be a continuation of the trend away from resolving contract disputes by litigation and arbitration, towards using quicker and cheaper procedures such as mediation, conciliation and adjudication.

Partnering contracts are a development advocated by promoters and contractors over recent years, with these objectives

- reducing tendering times and costs
- anticipating and avoiding disputes
- securing contractors' best staff
- reducing duplication of management and supervision
- achieving quicker start to construction.

There are a few model sets of conditions established in the UK and internationally for these contracts and more will be developed if the method continues to find favour.

In the UK payment to contractors for the achievement of milestone activities on a critical path programme is tending to replace admeasurement based on bills of quantities. This releases engineers and quantity surveyors from the costly drudgery of detailed measurement and enables them to concentrate more on planning, anticipating problems, managing the use of resources and controlling variations.

Training

At the time of publication there is a substantial skills shortage in the construction industry and investment in training is a characteristic of countries that achieve such substantial industrial growth. A much greater scale of training at all levels is required and needs to be sustained in the UK if it is to become more efficient and remain competitive in Europe and world markets.

To maintain an ability to provide economic and safe services, all engineers should keep up-to-date with the latest information through training and development and take the lead in using it. They should champion training and development for themselves and others, in order to continually improve professional knowledge, skills and attitudes, and they should show that this is being done.

Political and environmental requirements

Pollution

The 20th century has experienced an enormous growth in the world's population, and consequently in the demands for urbanisation, transport, power and other polluting activities. Responding to these demands is now subject to ecological and other questions on the potential effects of atmospheric and other pollution, such as

- gas, chemical and smoke emissions – mainly due to road vehicles, domestic heating systems, power stations and factories
- noise and vibration – due to roads, aircraft, railways, factories and construction
- polluted water and contaminated land – due to industrial discharges, fertilisers, sewage and waste treatment and disposal
- visual – all construction and transportation, waste ground and neglected structures.

Many people in the developed world may agree to try to reduce pollution. The developing world will have to continue these same polluting activities in order to maintain employment and services, let alone compete and develop. There is thus a major role for civil engineers in devising methods to satisfy these conflicting demands.

Sustainable development

As social and cost pressures grow to restrict the use of naturally available materials such as rock aggregates, engineers have to devise innovative economic methods for recycling old materials, making better use of existing materials and using waste from previous mining and other activities.

Whole-life costing

The application of whole-life cost-benefit analysis to decisions in project design is changing the evaluation and selection of potential infrastructure projects. Engineers have to master and apply these techniques if they are to advise on projects and manage their design.

The engineering profession

Unification

The need for engineering teams and project managers with multi-disciplinary expertise, together with professional unification, may well result in the blurring of traditional distinctions between various professional skills.

Unification of the UK Engineering Institutions will create a single organisation representing over 270 000 Chartered Engineers, the largest professional body in the country by far. This should have considerable influence with government, and not only change the way that engineers relate within the profession, but also improve the status and voice of the Chartered Engineer in society.

The need to establish of a single voice to represent all the UK professional organisations concerned with building, civil engineering and industrial construction has brought the Institution of Civil Engineers closer to its related institutions and other bodies. This closer working within construction is likely to continue, but become more complex with progressive integration of industrial and academic standards in Europe.

Continual improvement

Promoters increasingly expect that engineers will deliver their projects on time, at the right quality, lower cost and higher standards of health and safety. The public expect professional leadership to reduce the potential effects of new construction, changes to existing structures and demolition work on the environment. Organisations and individuals expect more success in competing for work. These expectations can be achieved only by more rapid and successful innovations in design, construction, contracts and management.

The need for continuous professional development will never be greater. It is a matter of survival to develop and use new ideas and technology successfully. Understanding commercial, legal, environmental and organisational factors and the use of value engineering and quality assurance concepts is increasingly expected of most engineers. They must gain expertise in these before others take the lead in them. There is also a great need for engineers to acquire and develop a wider range of managerial expertise and personal skills.[1] A first degree is not an end in itself; for most engineers it is the start to acquiring an increasingly broader range of expertise and skills.

Reference

1. *Management development in the construction industry – Guidelines for the professional engineer*, Thomas Telford, London, 1992

Appendix A
Model conditions of contract

Publications on UK practice and its problems are mostly written in terms of the use of various sets of 'model conditions of contract' for construction published by UK professional institutions, government departments and trade associations. These models are similar in their purpose of providing terms of contract to specify the responsibilities of the parties and the formal system of communications between them. They differ in the allocation of risks and in the detail of procedures and liabilities, and many clients add to them or have their own versions, but they are well known in construction in the UK and so probably influence clients' contract strategies. Listed here are the principal models which illustrate contract arrangements in practice. Many of these sets of terms are often called 'standards' in the UK, but 'models' is more accurate as they are not used uniformly.

ICE 7	*Conditions of Contract for Civil Engineering Works.* Institution of Civil Engineers, the Association of Consulting Engineers and the Federation of Civil Engineering Contractors, 7th edition (measurement edition), 2003 (ICE 7)
NEC	The NEC suite, published by Thomas Telford Limited (a wholly owned subsidiary of the Institution of Civil Engineers (ICE)):

The third edition (NEC3) was published in 2005 and now comprises the following:
Engineering and Construction Contract
Adjudicator's Contract
Engineering and Construction Short Contract
Engineering and Construction Subcontract
Framework Contract
Term Service Contract.

CECA *Form of Sub-contract* – for use with the ICE 6th edition. Civil Engineering Contractors Association, 1998

ICE Minor Works *Conditions of Contract for Minor Works.* Institution of Civil Engineers, 3rd edition, 2001

ICE Design and Construct *Conditions of Contract for Design and Construct.* Institution of Civil Engineers, 2nd edition, 2001

ICE cond. for Ground Investigation *Conditions of Contract for Ground Investigation.* Institution of Civil Engineers, Association of Consulting Engineers and Federation of Civil Engineering Contractors, 2nd edition, 2003

GC/Works/1 *GC/Works Edition 3 General Conditions of Contract for Building and Civil Engineering.* Published by HM Stationery Office: Lump Sum with Quantities version 1998; Lump Sum without Quantities version 1998; Single Stage Design & Build version 1998. (Notes: (i) Previous editions were rather different; (ii) Government departments in the UK are not required to use these models).

JCT 2005 Joint Contracts Tribunal *Standard Building Contract*, 2005. Published in alternative versions with and without quantities or approximate quantities

JCT 2005	*Design and Build Contract.* JCT
JCT 2005	*Intermediate Form of Building Contract.* JCT, 2005 (for smaller building projects)
JCT 2005	*Minor Works Contract (for small projects/ construction works)*
JCT 2005	*Major Project Construction Contract*
JCT 2005	*Construction Management Contract*
JCT 2005	*Management Building Contract*
JCT 2005	*Measured Term Contract*
JCT 2005	*Framework Agreement*
JCT 2005	*Prime Cost Contract*
IChemE Green Book	*Conditions of Contract for Complete Process Plants Suitable for Reimbursable Contracts.* Institution of Chemical Engineers, 3rd edition, 2002
IChemE Red Book[*]	*Conditions of Contract for Complete Process Plants Suitable for Lump Sum Contracts in the UK.* Institution of Chemical Engineers, 4th edition, 2001. (Though indicated in the title of the above model, payment is rarely in a lump sum for large projects.)
IChemE Yellow Book[†]	*Conditions of Subcontract for Process Plants.* Institution of Chemical Engineers, 3rd edition, 2003.

---

[*] FIDIC model conditions for international civil engineering contracts are also known as the 'Red Book'.
[†] FIDIC model conditions for international mechanical and electrical contracts have for long also been known as the 'Yellow Book'.

MF/1   *General Conditions of Contract*, model form MF/1,
               for home or overseas contracts – with erection. The
               Institution of Mechanical Engineers, the Institution
               of Electrical Engineers and the Association of
               Consulting Engineers, revision 4, 2000 ed.
               (replacing previous model forms A and B3).

MF/2   *General Conditions of Contract*, model form MF/2,
               for home or overseas contracts – without
               erection. The Institution of Mechanical
               Engineers, the Institution of Electrical Engineers
               and the Association of Consulting Engineers,
               revision 1, 1999 ed.

Professional services contracts

ACE   *Conditions of Engagement* – to form the basis for
               an agreement between client and consulting
               engineer for one or more stages of a project.
               Association of Consulting Engineers, revised 1995

APM   *Standard Terms for the Appointment of a Project
               Manager* – with a schedule of duties and
               responsibilities for the construction industry.
               Association of Project Managers, 1998

NEC   *Professional Services Contract*. 3rd edition, 2005

NEC   *Adjudicator's Contract*. 3rd edition, 2005

International construction

FIDIC New   *Conditions of Contract (International) for Building
Red Book   *and Engineering Works designed by the Employer*,
               Federation Internationale des Ingenieurs-Conseils
               (FIDIC), 1st edition, 1999. (This model has many
               similarities to the ICE 5 model conditions of
               contract, but also some differences.)

FIDIC New Yellow Book	*Conditions of Contract (International) for Plant and Design-Build for Electrical and Mechanical Plant, and for Building and Engineering Works designed by the Contractor.* Federation International des Ingenieurs-Conseils (FIDIC), 1st edition, 1999. (This model has many similarities to the IMechE/ IEE/ACE MF/1 model conditions of contract, but also some differences.)
FIDIC New Silver Book	*Conditions of Contract (International) for EPC and Turnkey projects.* Federation International des Ingenieurs-Conseils (FIDIC), 1st edition, 1999
FIDIC New White Book	*Client/Consultant Model Services Agreement* Federation International des Ingenieurs-Conseils (FIDIC), 4th edition, 2006
UNECE	*General Conditions for the Supply of Plant and Machinery for Export.* United Nations Economic Commission for Europe (UNECE) (alternatives for supply with and without erection)
EDF	*General Rules and Procedures for Works, Supply and Services Contracts Financed by the European Development Fund.* European Development Fund, 1990 *General Conditions for Works Contracts.* European Development Fund, 2006

Definitions used in model conditions of contract

Table A.1 lists some of the equivalent definitions used in the more common model sets of conditions of contract for larger projects. These are approximate equivalents, dependent on their definition in the particular conditions of contract.

The numbers indicate the clauses where these words are defined or first used in the set of conditions. For the NEC the numbers refer to the core clauses except where Option A is indicated.

Table A.1 Definitions used in model conditions of contract

ICE 7 and FIDIC (Civil engineering)	NEC	GC/Works/1	JCT	IChemE (Red and Green Books)	MF/1 and MF/2
Employer	Employer	Employer	Employer	Purchaser	Purchaser
Contractor	Contractor	Contractor 1	Contractor	Contractor	Contractor
Engineer	Project Manager Adjudicator Supervisor	PM Adjudicator	Architect or Contract Administrator	Project Manager	Engineer
Engineer's representative (ICE) Engineer's assistants[1] (FIDIC)		Clerk of Works or Resident Engineer[2]	Clerk of Works	Project Manager's representative	Engineer's representative
Contractor's agent (ICE) Contractor's representative[3]		Contractor's agent	Person-in-charge	Site Manager	Contractor's representative
Tender Total (ICE)[4]	Prices	Contract sum	Contract sum	Contract price	Contract price
Contractor's Equipment (ICE)	Equipment	Contractor's plant	Contractor's plant	Contractor's equipment 1	Contractor's equipment

Table A.1 Continued

ICE 7 and FIDIC (Civil engineering)	NEC	GC/Works/1	JCT	IChemE (Red and Green Books)	MF/1 and MF/2
Interim Certificate (ICE)[5]	Payment certificate	Advances on account	Interim certificate	Interim statement	Interim certificate
Final certificate	Payment certificate after completion of the whole of the works	Final certificate for payment	Final certificate	Final certificate	Final certificate of payment
Certificate of substantial completion[6]	Completion	Certificate of completion of the works	Practical completion certificate	Take-over certificate	Taking-over certificate
Defects correction period (ICE)[7]	Defect correction period	Maintenance period	Rectification period	Defects liability period	Defects liability period
Defects correction certificate[8]	Defects certificate		Certification of completion of making good	Final certificate	Final certificate of payment

[1] FIDIC

[2] Are optional appointments in relation to performing the PM's powers in relation to assessing the quality of the Works.

[3] Representative (FIDIC)

[4] Contract Price (FIDIC)

[5] (ICE) Interim Payment Certificate (FIDIC)

[6] (ICE) Taking Over Certificate (FIDIC)

[7] (ICE) Defects Notification Period (FIDIC)

[8] (ICE) Performance Certificate (FIDIC)

Appendix B
Bibliography

This list should be used selectively, depending upon individual needs and future interests. No one is expected to read all of these publications. On some subjects there are alternatives. These and any new publications since this list was compiled may be available in the ICE's or other libraries and are usually on sale at the ICE's bookshop.

The promotion of projects

GRIFFITH A. and KING A. *Best Practice Tendering for Design and Build Projects*, Thomas Telford, London, 2003

KAMARA J. M., ANUMBA C. J. and EVBUOMWAN N. F. *Capturing Client Requirements in Construction Projects*, Thomas Telford, London, 2002

HAMILTON A. *Managing Projects for Success*, Thomas Telford, London, 2001

HOFFMAN S. L. L. *The Law and Business of International Project Finance*, Kluwer Law International, London, 2nd edition, 2001

Risk management

PERRY P. *Risk Assessment: Questions and Answers*, Thomas Telford, London, 2003

ACTURIAL PROFESSION and INSTITUTION of CIVIL ENGINEERS. Risk Analysis and Management for Projects, Thomas Telford, London, 2nd edition, 2005

OFFICE OF GOVERNMENT COMMERCE. *Risk and Value Management*, 2007, available http://www.ogc.gov.uk/ppm_documents_construction

150

Responsibilities for a project

HAMILTON A. *Handbook of Project Management Procedures*, Thomas Telford, London, 2004

Design

Design guides – series being published by Thomas Telford, London
AUSTIN S., THORPE A., HAMMOND J. and MURRAY M. *Design Chains*, Thomas Telford, London, 2001

Organisation

NAOUM S. *People and Organisational Management in Construction*, Thomas Telford, London, 2001
BENNETT J. and JAYES S. *Trusting the Team*, Thomas Telford, London, 1995
CORNICK T. and MATHER J. *Construction Project Teams*, Thomas Telford, London, 1999
CONSTRUCTION INDUSTRY COUNCIL. *The procurement of professional services*, London, 1993

Project management

MORRIS P. *The Management of Projects*, Thomas Telford, London, 1997
LEVY S. M. *Project Management in Construction*, McGraw-Hill, Maidenhead, 2006
LOCK D. *Project management*, Gower Publishing Co, London, 6th edition, 1996
CORRIE R. K. (ed.) *Project evaluation*, Thomas Telford, London, 1990

Health, Safety and Welfare

JOYCE R. *The CDM Regulations 2007 Explained*, Thomas Telford, London, 2007
PERRY P. *Health and Safety: Questions and Answers*, Thomas Telford, London, 2003
PERRY P. *CDM: Questions and Answers*, Thomas Telford, London, 2nd edition, 2002

Contract Strategy

MURDOCH J. and HUGHES W. *Construction Contracts: Law and Management*, Taylor and Francis, London, 2007

CURTIS B. *et al. Roles, responsibilities and risks in management contracting*, Publication SP81, Construction Industry Research & Information Association, London, 1991

HODGSON G. J. Design and build – Effects of contractor design on highway schemes, *Civil Engineering. Proc. Instn Civ. Engrs*, 1995, 108, pp. 64–76

LATHAM SIR MICHAEL. *Constructing the team*, report of the joint government industry review of procurement and contractual arrangements in the United Kingdom construction industry, HMSO, London, 1994

WRIGHT D. *An engineer's guide to the model forms of conditions of contract for process plant*, Institution of Chemical Engineers, Rugby, 1994

Contract Management

LIEBING R. W. *Construction Contract Administration*, Prentice Hall, London, 1997

LOFTUS J. *Project Management of Multiple Projects and Contracts*, Thomas Telford, London, 1999

ATKINSON, A. V. *Civil engineering contract administration*, Stanley Thornes Publications, 2nd edition, 1992

EDWARDS L. J., LORD G. and MADGE P. *Civil engineering insurance and bonding*, Thomas Telford, London, 2nd edition, 1996

INSTITUTION OF CIVIL ENGINEERS. *Guidance on the preparation, submission and consideration of tenders for civil engineering contracts*, London, 1983

UFF J. F. *Construction law*, Sweet & Maxwell, London, 6th edition, 1994

WEARNE S. H. (ed.), *Civil engineering contracts*, Thomas Telford, London, 1989

Contract Supervision

COOKE B. and WILLIAMS P. *Construction Planning, Programming and Control*, Wiley-Blackwell, London, 2004

Construction Management

Cox A., Ireland P. and Townsend M. *Managing in Construction Supply Chains and Markets*, Thomas Telford, London, 2006

Harris F. and McCaffer R. *Modern construction management*, Blackwell Scientific Publications, Oxford, 4th edition, 1995

Shaughnessy H. (ed.) *Collaboration management – New project and partnering techniques*, John Wiley & Sons, New York, 1994

Works construction guides published by Thomas Telford, London

Site Investigation Steering Group. *Site investigation in construction series*, Institution of Civil Engineers, London, 1993

The Green Book: Appraisal and Evaluation in Central Government, Treasury Guidance, TSO. Available http://www.hm-treasury.gov.uk/economic_data_and_tools/greenbook/data_greenbook_index.cfm

Best Practice OGC Gateway Reviews, Office of Government Commerce, TSO

Sustainable Construction Strategy Report, Department of Trade and Industry, 2006, TSO

Building for the Future: Sustainable construction and refurbishments on the Government Estate, National Audit Office, 2007, TSO

Cost Control

Bower D. *Management of Procurement*, Thomas Telford, London, 2003

Merna A. *Financing Infrastructure Projects*, Thomas Telford, London, 2002

McCaffer R. C. and Baldwin A. N. *Estimating and tendering for civil engineering works*, Blackwell Scientific Publications, Oxford, 2nd edition, 1991

Smith N. J. *Project cost estimating*, Thomas Telford, London, 1995

Overseas Projects

Howes R. and Tah J. H. M. *Strategic Management Applied to International Construction*, Thomas Telford, London, 2003

Bennett J. *International Construction Project Management: General Theory and Practice*, Butterworth-Heinemann Ltd, Oxford, 1991

LORAINE R. K. *Construction management in developing countries*, Thomas Telford, London, 1991

Professional Duties

MacDONALD STEELS H. *Successful Professional Reviews for Civil Engineers*, Thomas Telford, London, 2nd edition, 2006

SCOTT W. *Communication for Professional Engineers*, Thomas Telford, London, 2nd edition, 1997

ARMSTRONG J., DIXON R. and ROBINSON S. *The Decision Makers: Ethics for Engineers*, Thomas Telford, London, 1999

MacDONALD STEELS H. *Effective training for civil engineers*, Thomas Telford, London, 1994

CLAYTON R. (ed.). *Guide to consultancy*, Institution of Chemical Engineers, Rugby, 2nd edition, 1995

Appendix C

Useful addresses and websites

Institution of Civil Engineers
(ICE)
One Great George Street
Westminster
London
SW1P 3AA
www.ice.org.uk

Association for Consultancy and
Engineering (ACE)
Alliance House
12 Caxton Street
London
SW1H 0QL
www.acenet.co.uk

Association for Project
Management (APM)
150 West Wycombe Road
High Wycombe
Buckinghamshire
HP12 3AE
www.apm.org.uk

Construction Confederation
(formally Building Employers
Confederation)
55 Tufton Street
Westminster
London
SW1P 3QL
www.thecc.org

Chartered Institute of Building
(CIOB)
Englemere
Kings Ride
Ascot
Berkshire
SL5 7TB
www.ciob.org.uk

Construction Industry Council
(CIC)
26 Store Street
London
WC1E 7BT
www.cic.org.uk

Construction Industry Research
& Information Association
(CIRIA)
Classic House
174–180 Old Street
London
EC1V 9BP
www.ciria.org.uk

ICES
Dominion House
Sibson Road
Sale
Cheshire
M33 7PP
United Kingdom
www.ices.org.uk

Institution of Chemical
Engineers (IChemE)
London office
1 Portland Place
London
W1B 1PN
Rugby office
Davis Building
165–189 Railway Terrace
Rugby
CV21 3HQ
www.icheme.org

Chartered Institute of
Arbitrators
International Arbitration and
Mediation Centre

12 Bloomsbury Square
London
WC1A 2LP
www.arbitrators.org

Chartered Institute of
Purchasing & Supply (CIPS)
Easton House
Easton on the Hill
Stamford
Lines
PE6 3NZ
www.cips.org

Institution of Engineering
Technology (IET)
Michael Faraday House
Six Hills Way
Stevenage
SG1 2AY
www.iet.org

Royal Institute of British
Architects (RIBA)
66 Portland Place
London
W1B 1AD
www.architecture.com

Thomas Telford Ltd
Thomas Telford House
1 Heron Quay
London
E14 4JD
www.thomastelford.com

Appendix D

Glossary

Admeasurement – apportioning of quantities or costs. See also *Remeasurement, Valuation, Bills of quantity*

Adjudicator – in the NEC system and under the statutory requirement for adjudication under the Housing Grants, Construction and Regeneration Act 1996, the person appointed to give a decision on a dispute between the parties to the contract

Agent – in civil engineering in the UK 'Agent' is traditionally the title of a contractor's representative on a site. Many now have the title 'Project Manager' (see below)

Bid – an offer to enter into a contract; such an offer by a contractor to construct a project. Also called a 'Tender'

Bills of quantity (BoQ) – a list of the items and quantities of delivered work to be done for the Promoter under a contract, for instance a quantity of concrete placed to a specified quality. An equivalent in some industrial contracts is called a schedule of measured work. See also *Schedule of rates, Unit rate*. In the traditional arrangements a BoQ is normally issued with an invitation to contractors to tender for a project, with a specification and the tender drawings, and the contractors insert their 'rates' (prices per unit quantity) for each item. The rates in the contractors' tenders can then be compared item by item. In what are called 'admeasurement' contracts, payment to a contractor is based on these rates × the final quantities of work done

CDM – the Construction (Design & Management) Regulations, 2007. UK legal requirements implementing European Commission Directive 92/57/EEC

157

CDM *Co-ordinator* – a competent person appointed by the Promoter (client) under CDM whose role is to ensure the competence of designers and contractors before they are appointed. One of the main duties of the CDM C-oordinator is to ensure that a Health and Safety Plan for the project is prepared before construction work starts

Client – see *Promoter*

Conditions of contract – the 'conditions' of a contract means all the important terms of that contract. In civil engineering the words 'conditions of contract' are used to mean sets of contract terms on general matters likely to be required in all contracts for a class of projects. They define the words used, the responsibilities of the parties, procedures, liabilities for damage, injuries, mistakes or failures of contractor or sub-contractors, delays, changes in legislation such as taxation, frustration of contract and termination. They are designed to be used with a specification, drawings, schedules and other documents which state the particular terms of a contract. They are also called 'general conditions of contract', 'model forms' or 'standard forms'. See also *Terms of a contract*

Construction Management – is used with a special meaning in the USA and the UK to mean the employment by a Promoter of a Construction, see *Management Contractor*, Manager to help plan, define and coordinate design and construction, and supervise construction by 'Trade Contractors' (see below). The Trade Contractors are engaged by the Promoter but act under the direction of the Construction Manager

Construction Manager – a person or organisation employed by the Promoter to help plan, define and coordinate design and construction and to direct and supervise construction by other 'Trade Contractors' – see *Trade Contractor*

Consultants – professional advisers on studies, projects, design, management, techniques and technical or other problems

Consulting engineers – consultants who also design projects and supervise construction, usually in firms of partners and supporting staff

Contract – an agreement enforceable at law

Contractor – in general a supplier of services; in civil engineering usually a company which undertakes the construction of part or all a project, and for some projects also undertakes design or other services

The Contractor – the party to a contract, as distinct from any contractor. The word 'the' is important in English practice as identifying the particular contractor who has entered into a contract for a project. The Contractor may be a joint venture of two or more companies

Cost plus – see *Reimbursable contract*

Defects correction Period – see *Retention*

Defects liability period (sometimes called maintenance period) – see *Retention*

Direct labour – a Promoter's own employees employed on construction, sometimes under the internal equivalent of a contract, otherwise as a service department

Early Contractor Involvement – used as an expression to denote a non traditional procurement route, where a contractor's skills are introduced early into a project to bring design, 'buildability' and cost efficiencies to the pre construction phase

The Employer – see *Promoter*

The Engineer – a person named in some forms of contract to be responsible for administering that contract, particularly in contracts for construction or for the supply and installation of equipment. Also called Project Manager under the NEC Form of Contract

The Engineer's Representative – in ICE contracts the formal title for the Engineer's representative on site, often called 'the Resident Engineer' – see also *Supervisor*

Equipment – machines, services and other systems. 'Contractor's equipment' is defined in some civil engineering contracts to mean things used by the contractor to construct the works, but not materials or other things forming part of the permanent works – see also *Plant*

Expression of Interest – where a contractor, consultant or other supplier of services informs a promoter that it is interested in competing for work

Feasibility studies – investigation of possible designs and estimating their costs to provide the basis for deciding whether to proceed with a proposed project

Firm price – varies in its meaning, but is often used to indicate that a tendered price is offered only for a stated period and is not a commitment if it is not accepted within that period – see below

Fixed price – usually means that a tender price will not be subject to escalation, but it may mean that there is no variations term

NB Like some other words used in contract management, 'fixed price' has no fixed meaning and 'firm price' no firm meaning. What matters in each contract is whether its terms of payment include provisions for changes to the contract price, and what the governing law permits

Framework – a contract commissioned by a promoter for the provision of works or services of indeterminate nature. When specific works requirements are identified, a contractor or consultant who has successfully tendered a framework contract will be appointed to undertake the works

Lump sum payment – used in engineering and construction to mean that a contractor is paid on completing a major stage of work, for instance on handing over a section of a project. Strictly it means payment in a single lump. In practice 'lump sum' is used to mean that the amount to be paid is fixed, based on the contractor's tender price but perhaps subject to contract price adjustment

Main contractor – similar to a general contractor: a contractor who undertakes the construction of all or nearly all a project, but usually in turn sub-letting parts of his work to specialist or trades contractors and others as sub-contractors but as distinct from a 'Management' or 'Managing' Contractor and/or a Construction Manager

Maintenance contract – a contract to maintain or operate the works upon completion of the construction of those works

Management or Managing Contractor – a contractor employed by the Promoter to help plan, define and coordinate design and construction and to direct and supervise construction by other 'Works Contractors' – see *Works Contractor*. The Works Contractors are engaged by, and act under the direction of, the Management Contractor

Measurement – calculation of quantities of work for a BoQ or for payment – see also *Remeasurement*

Milestone and *planned progress payment* – payment to a contractor in a series of lump sums each paid upon his achieving a 'milestone' – meaning a defined stage of progress. Use of the word milestone usually means that payment is based upon progress in completing what the Promoter wants. Payment based upon achieving defined percentages of a contractor's programme of activities is also known as a 'planned payment' scheme

Nominated sub-contractor – a sub-contractor usually for specialist work who is chosen by the Promoter or the Engineer rather than by the main contractor, but is then employed by the main contractor

Partnering – collaborative management of a contract by promoter and contractor to share risks and rewards

Permanent works – the works to be constructed and handed over to the Promoter

Plant – traditionally 'Contractor's plant' is defined in civil engineering contracts to mean things used by the contractor to construct the works, but not materials or other things forming part of the permanent works – see also *Equipment*

Pre-qualification – the process of inviting the consultants, contractors or sub-contractors who are interested in tendering for work first to submit information on their relevant experience, performance, capacity, resources, systems and procedures, and from this information assessing which are qualified to be invited to tender – see also *Qualification*

Principal contractor – the contractor appointed under the CDM Regulations (see above) to be responsible for ensuring compliance with the health and safety plan by all contractors and individuals on construction – usually the main or largest contractor on a site

Private Finance Initiative – see description in Chapters 2 and 4

Project – any new structure, system or facility, or the alteration, renewal, replacement, substantial maintenance or removal of an existing one

161

Project Engineer – the title often used in promoters' and consultants' organisations for the role of an engineer responsible for leading and coordinating design and other work for a project. In some cases this title is used where the role is actually the greater one of Project Manager as described below

Project management – sometimes used with the special meaning that the Promoter employs an independent professional project manager to help plan, define and coordinate the work of consultants and contractors to design and construct a project

Project Manager/Promoter's Manager – the title increasingly used in promoters', contractors' and consultants' organisations for the role of manager of the development and implementation of a project. The role may have other titles, such as 'Project Director' for a large project. For smaller projects the role is not necessarily a separate job. The term Project Manager is also used in various forms of contract to describe the role undertaken by the Contract Administrator see *The Engineer* and *Construction Manager*

The Promoter – the 'client' for a project, the individual or organisation that initiates a project and obtains the funds for it. In some contracts the Promoter is called 'the Employer', in others 'the Owner', 'the Purchaser' or 'the Principal'

Public Private Partnership – see description in Chapters 2 and 4

Qualification – commonly used to mean an accomplishment or attribute which is recognised as making a person or an organisation fit to undertake a specified role or function – see also *Pre-qualification* – also used to mean that a tender includes reservations or statements made to limit liabilities if that tenderer is given the contract

Reimbursable contract – a contract under which a promoter pays ('reimburses') all a contractor's actual costs of all his employees on the contract ('payroll burden') and of materials, equipment and payments to sub-contractors, plus usually a fixed sum or percentage for management, financing, overheads and profit. This is often called a 'cost plus' contract

Remeasurement – calculation of the actual quantities of work ordered on the contractor in order to certify the payment due to a contractor. Remeasurement is also known as 'measure-and-value'

Retention (retention money) – a part of the payment due to a
contractor for progress with the work which is not paid until he
has discharged liabilities to remedy defects for a period after the
taking over or acceptance of the works by the Promoter. In
some contracts this period is called the 'Defects correction
period' or 'Defects liability period'. It is typically 12 months –
tends to be longer these days
'Defects correction period' varies under NEC depending on date of
notification of defect

Schedule of rates – what is called the 'schedule of rates' in some
contracts is very similar to a bill of quantities in form and purpose.
Contractors when bidding are asked to state rates per unit of items
on the basis of indications of possible total quantities in a defined
period or within a limit of say ±15% variation of these quantities.
In other examples, the rates are to be the basis of payment for any
quantity of an item which is ordered at any time, for instance in
term contracts for maintenance and minor construction work
and 'dayworks' schedules included in a traditional civil engineering contract

Specialist contractors – contractors who limit their work to selected
types of work, e.g. piling or building services. They are often
sub-contractors to general contractors. See also *Trades contractors*
and *Works Contractors*

Sub-contractor – a contractor employed by a main contractor to carry
out part of a project. A sub-contractor is not in contract with the
promoter

Sub-letting – employment of a sub-contractor by a main contractor

Supervisor – used in the NEC system to mean the person appointed
to check that the Works are constructed in accordance with the
contract – see also *Engineer's Representative*

Temporary works – items built to facilitate the construction of the Works

Tender – an offer to enter into a contract, such an offer by a
contractor to construct a project. Also called a 'bid'

Term contract – a system in which a Promoter invites several contractors to give prices for typical work which is to be carried out if and
when ordered at any time during an agreed period ('the term'),

usually based upon descriptions of types of work which may be ordered but without quantities being known in advance

Terms of a contract — all the obligations and rights agreed between the parties, plus any terms implied by law

Trade Contractor — a contractor employed by a Promoter under a Construction Management arrangement, see also *Construction Management*

Turnkey contract — a contract in which the contractor is responsible for the design, supply, construction and commissioning of a complete structure, factory or process plant

Unit rates basis of payment — payment at a fixed price ('rate') per unit of work done. In UK civil engineering 'admeasurement' contracts the predicted total amounts of work are usually stated item by item in a bill of quantities (BoQ) and the contractor is paid the rate for each × actual amount of work done

Valuation — in building and civil engineering contracts in the UK the process of calculating a payment due to the contractor

Value engineering — an analytical technique for questioning whether the scope of a design and the quality of proposed materials will achieve a project's objectives at minimum cost

Variation — a change to the quantity, quality or timing of the works which is ordered by the Promoter's representative under a term of a contract

Working Rule Agreement — terms of employment agreed between one or more trades unions and representatives of the employers of the members of those unions

The Works — what a contractor has undertaken to provide or do for a promoter — consisting of the work to be carried out, goods, materials and services to be supplied, and the liabilities, obligations and risks to be taken by that contractor. It may not be all of a project, depending upon what is specified in a contract

For other definitions see Blockley D. *The new Penguin dictionary of civil engineering*, Penguin Books, 2005, and in a contract the definitions particular to that contract

Index

Page numbers in *italics* refer to Figures.
Individual job titles are given with Initial Capital Letters.